U.S. Department of
Transportation

Federal Railroad
Administration

Track Buckling Prevention: Theory, Safety Concepts, and Applications

Office of Research
and Development
Washington, D.C. 20590

Track Systems Safety

DOT/FRA/ORD-13/16

Final Report
March 2013

This document is available to the
public through the National Technical
Information Service, Springfield, VA 22161.
This document is also available on the FRA
Web site at www.fra.dot.gov.

REPORT DOCUMENTATION PAGE

Form Approved
OMB No. 0704-0188

1 AGENCY USE ONLY *(Leave blank)*	2 REPORT DATE	3 REPORT TYPE AND DATES COVERED
	March 2013	Final Report

4 TITLE AND SUBTITLE	5 FUNDING NUMBERS
Track Buckling Prevention: Theory, Safety Concepts, and Applications	RR219/R2011; RR319/RR3011; RR419/AB019

6 AUTHOR(S)

Andrew Kish and Gopal Samavedam*

7 PERFORMING ORGANIZATION NAME(S) AND ADDRESS(ES)	8 PERFORMING ORGANIZATION REPORT NUMBER
U.S. Department of Transportation Research and Innovative Technology Administration John A. Volpe National Transportation Systems Center Cambridge, MA 02142	DDTS.020104

9 SPONSORING/MONITORING AGENCY NAME(S) AND ADDRESS(ES)	10 SPONSORING/MONITORING AGENCY REPORT NUMBER
U.S. Department of Transportation Federal Railroad Administration 1200 New Jersey Ave., SW Washington, DC 20590	DOT/ FRA/ORD-13/16

11 SUPPLEMENTARY NOTES

*Foster-Miller, Inc.
350 Second Avenue
Waltham, MA 02451-1196

12a DISTRIBUTION/AVAILABILITY STATEMENT	12b DISTRIBUTION CODE
This document is available to the public through the National Technical Information Service, Springfield, VA 22161 and on the FRA Web site at www.fra.dot.gov.	

13 ABSTRACT *(Maximum 200 words)*

This report is a part of the John A. Volpe National Transportation Systems Center's Track Stability Research Program for the Federal Railroad Administration on thermal buckling of continuous welded rail (CWR) track and its prevention. Presented in this report are the developments of theoretical results and the development and application of the CWR-SAFE computer software model for prediction of CWR track buckling strength. This comprehensive predictive model encompasses several different modules designed to perform both deterministic and probabilistic buckling analyses, based on the dynamic buckling theory previously validated by tests, and predicts safe limits for buckling prevention. The model accounts for all the important parameters influencing track buckling, such as rail size, curvature, lateral resistance, tie-ballast friction, fastener torsional and longitudinal resistances, track vertical stiffness, misalignment amplitude and wavelength, and vehicle parameters. Applications of the model are demonstrated through analyses of parametric sensitivity, development of buckling safety limits in terms of safe and critical temperatures, and evaluation of annual probability of buckling occurrences for typical CWR line segments. The report also presents techniques to determine the input parameters for CWR-SAFE application and a practical methodology for CWR track safety monitoring. A risk-based approach is proposed to provide more flexibility to the industry in achieving a minimum number of annual buckles in a given territory and to provide science-based guidelines for improved slow order policies when operating at elevated rail temperatures.

14 SUBJECT TERMS	15 NUMBER OF PAGES
Track Buckling, Lateral Stability, CWR, Buckling Prevention, Buckling Models, CWR Safety and Maintenance, Neutral Temperature, Rail Temperature	168
	16 PRICE CODE

17 SECURITY CLASSIFICATION OF REPORT	18 SECURITY CLASSIFICATION OF THIS PAGE	19 SECURITY CLASSIFICATION OF ABSTRACT	20 LIMITATION OF ABSTRACT
Unclassified	Unclassified	Unclassified	

NSN 7540-01-280-5500

Standard Form 298 (Rev. 2-9)
Prescribed by ANSI Std. 239-18
298-102

METRIC/ENGLISH CONVERSION FACTORS

ENGLISH TO METRIC

LENGTH (APPROXIMATE)

1 inch (in) = 2.5 centimeters (cm)

1 foot (ft) = 30 centimeters (cm)

1 yard (yd) = 0.9 meter (m)

1 mile (mi) = 1.6 kilometers (km)

AREA (APPROXIMATE)

1 square inch (sq in, in^2) = 6.5 square centimeters (cm^2)

1 square foot (sq ft, ft^2) = 0.09 square meter (m^2)

1 square yard (sq yd, yd^2) = 0.8 square meter (m^2)

1 square mile (sq mi, mi^2) = 2.6 square kilometers (km^2)

1 acre = 0.4 hectare (he) = 4,000 square meters (m^2)

MASS - WEIGHT (APPROXIMATE)

1 ounce (oz) = 28 grams (gm)

1 pound (lb) = 0.45 kilogram (kg)

1 short ton = 2,000 pounds (lb) = 0.9 tonne (t)

VOLUME (APPROXIMATE)

1 teaspoon (tsp) = 5 milliliters (ml)

1 tablespoon (tbsp) = 15 milliliters (ml)

1 fluid ounce (fl oz) = 30 milliliters (ml)

1 cup (c) = 0.24 liter (l)

1 pint (pt) = 0.47 liter (l)

1 quart (qt) = 0.96 liter (l)

1 gallon (gal) = 3.8 liters (l)

1 cubic foot (cu ft, ft^3) = 0.03 cubic meter (m^3)

1 cubic yard (cu yd, yd^3) = 0.76 cubic meter (m^3)

TEMPERATURE (EXACT)

$[(x-32)(5/9)]$ °F = y °C

METRIC TO ENGLISH

LENGTH (APPROXIMATE)

1 millimeter (mm) = 0.04 inch (in)

1 centimeter (cm) = 0.4 inch (in)

1 meter (m) = 3.3 feet (ft)

1 meter (m) = 1.1 yards (yd)

1 kilometer (km) = 0.6 mile (mi)

AREA (APPROXIMATE)

1 square centimeter (cm^2) = 0.16 square inch (sq in, in^2)

1 square meter (m^2) = 1.2 square yards (sq yd, yd^2)

1 square kilometer (km^2) = 0.4 square mile (sq mi, mi^2)

10,000 square meters (m^2) = 1 hectare (ha) = 2.5 acres

MASS - WEIGHT (APPROXIMATE)

1 gram (gm) = 0.036 ounce (oz)

1 kilogram (kg) = 2.2 pounds (lb)

1 tonne (t) = 1,000 kilograms (kg)

= 1.1 short tons

VOLUME (APPROXIMATE)

1 milliliter (ml) = 0.03 fluid ounce (fl oz)

1 liter (l) = 2.1 pints (pt)

1 liter (l) = 1.06 quarts (qt)

1 liter (l) = 0.26 gallon (gal)

1 cubic meter (m^3) = 36 cubic feet (cu ft, ft^3)

1 cubic meter (m^3) = 1.3 cubic yards (cu yd, yd^3)

TEMPERATURE (EXACT)

$[(9/5)y + 32]$ °C = x °F

QUICK INCH - CENTIMETER LENGTH CONVERSION

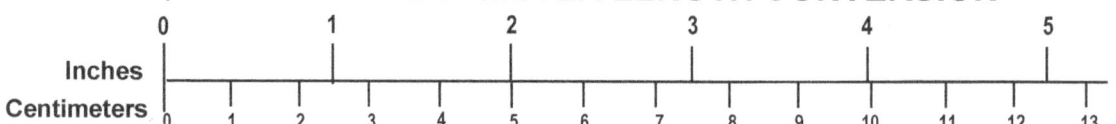

QUICK FAHRENHEIT - CELSIUS TEMPERATURE CONVERSIC

For more exact and or other conversion factors, see NIST Miscellaneous Publication 286, Units of Weights and Measures. Price $2.50 SD Catalog No. C13 10286

Updated 6/17/98

PREFACE

This report provides a current state-of-the-art understanding of continuous welded rail (CWR) response and behavior and its impact on track safety and performance. The report is based on U.S. research over the past 20 years, including the development of versatile computer analysis software called CWR-SAFE. CWR-SAFE is a product of Volpe Center, Foster-Miller, Inc. (FMI), and Federal Railroad Administration (FRA) research under the latter's Track and Structures Program initiatives in line with the FRA's Five-Year Strategic Plan for Railroad Research, Development, and Demonstration (2002). The work reported herein was performed with the support of FMI under Contract DTRS-57-99-D-0003, Task Order #14, under the technical direction of Dr. Andrew Kish of the US Department of Transportation's Volpe Center and FMI's Project Manager and Principal Investigator, Dr. Gopal Samavedam. The FRA sponsor was Dr. Magdy El-Sibaie of the Office of Research & Development, whose assistance and support is greatly acknowledged. Additional thanks also go to Mr. Wesley Mui of the Volpe Center for performing the CWR-SAFE parametric studies.

The report is intended to synthesize into one reference volume the various CWR relevant research results obtained at different periods during a 20-year program duration. The report covers the following research items performed from 1984 to 2004:

- Development of vehicle induced (dynamic) buckling theory for the determination of critical conditions leading to track buckling

- Identification and determination of track and vehicle parameters in the theory

- Development of a versatile computer software (CWR-SAFE) for application of the theory

- Parametric analyses of track buckling

- Development of buckling safety limits

- Tests to validate the theory and safety limits

- Determination of annual probability of buckling estimates for typical CWR line segments based on probabilistic analysis

- Methodologies to implement and monitor the safety limits for improved buckling prevention

- Development of science-based guidelines for slow order rationale for vehicle operations at high rail temperatures

CONTENTS

FIGURES

TABLES

Executive Summary

This report provides a comprehensive review of the research performed in the area of continuous welded rail (CWR) track safety and performance culminating in the development of CWR-SAFE computer analysis software under the sponsorship of the Federal Railroad Administration (FRA). The report also attempts to synthesize the various research items performed between 1984 and 2004 and compiles the results into one reference volume addressing the following key areas:

- Development of vehicle-induced (dynamic) buckling theory for the determination of critical conditions leading to track buckling

- Identification and characterization of track and vehicle parameters in the theory

- Development of the CWR-SAFE computer software for application of the theory

- Parametric analyses of track buckling

- Development of buckling safety limits

- Tests to validate the theory and safety limits

- Methodologies to implement and monitor the safety limits for improved buckling prevention

- Determination of probabilistic approaches to buckling evaluations

- Applications of probabilistic models to risk-based safety and maintenance issues

The theory embodied in CWR-SAFE includes both deterministic and probabilistic approaches; the latter was also used to develop risk-based safety limits for rail vehicle operations at elevated rail temperatures, including mitigation measures, such as risk index-based slow-order procedures.

Theoretical Aspects

The report reviews and presents the dynamic buckling theory for CWR track buckling evaluations. Track buckling under a moving train is found to be more prevalent than static buckling (i.e., without any train action). This is attributed to resistance loss resulting from uplift tendency of the track between the trucks of cars. A number of parameters including lateral resistance, longitudinal resistance, torsional resistance, rail section properties, track modulus, and vehicle parameters (such as truck center spacing and axle loads) have been identified as the key factors governing track buckling response. With large deflection theory, the appropriate nonlinear differential equations governing lateral deflection in the buckled zone and the longitudinal displacement in the adjoining zones are formulated. A Fourier series technique has been developed for a fast converging solution of the differential equations. The solution typically represents a nonlinear stability relationship between the lateral deflection and the rail temperature increase, which identifies two salient temperature points. One is the upper critical temperature, T_{Bmax}, at which the CWR track buckles explosively with no external energy. The other is T_{Bmin}, the lower critical temperature, at which the track can buckle with some external disturbance such as due to the moving train. The actual buckling temperature regime is between these two temperatures. In some instances (defined as progressive buckling), such a regime does not exist, in which case the critical temperature is defined as T_P (a temperature at which

1

increased lateral displacements begin to occur with small increases in rail temperature). Such progressive buckling, typified by a much slower or more gradual increase in lateral alignment, usually occurs at tracks with weak lateral resistance.

A number of full-scale static and dynamic buckling tests conducted at the Transportation Technology Center (TTC) validated the theory. In these tests, the track was instrumented to monitor rail longitudinal force and to measure uplift; then, the rails were artificially heated to induce the thermal load, and various buckling characteristics (such as explosive and progressive) were simulated on tangent and curved test segments. This report briefing summarizes some of the salient test results.

Buckling Parameters Characterization

Knowledge of key parameters, their characteristics, and their determination are very important aspects of CWR safety and performance. Parametric studies have identified that the peak value of the resistance influences the upper critical temperature, whereas the steady or limiting value influences the lower critical temperature. Nonlinear characteristic were determined in the field for various wood and concrete tie track and are described in the report. Also, by loading the ties vertically, the friction coefficient between the ballast and the tie can also be determined. This loaded resistance is also an important parameter affecting the buckling strength of CWR tracks because it influences the uplift regime, which controls the buckling process. The report also describes measurement techniques used to determine the other resistance parameters of longitudinal and torsional resistances.

Determination of neutral temperature (or the rail's stress free temperature), which controls the longitudinal force in the rail, has been a challenging research problem to date. The Rail Uplift Technique has been developed as a means of determining the rail force and the neutral temperature, in addition to the use of strain gages. The latter requires rail cutting for zero reference condition determination. The rail uplift technique requires unfastening the rail from the ties and the application of an uplift force to produce a vertical deflection. Various neutral temperature tests have identified many causes for its variation from the initial installation condition, including excessive rail creep because of ineffective fasteners/anchors, inadequate readjustment in conjunction with rail repair and destressing, curve pull-in/pull-out movement, and various maintenance practices, such as curve realignment, tie renewal, and surfacing. Some of these tests and analyses of buckled track data indicate occurrences of neutral temperature reductions down to the 50–60 °F range from initially installed values of 90–110 °F, and buckling analyses confirm that such low values can be highly instrumental in producing buckling prone conditions.

Computer Programs

This report discusses the computer program CWR-SAFE, which consists of three modules, CWR-BUCKLE, CWR-INDY, and CWR-RISK, developed under this research effort to evaluate buckling strength and perform safety analysis of CWR track. The program and all the modules operate in the Microsoft Windows environment and come with easily accessible User's Guides in the software's Self-Help Menu.

CWR-BUCKLE is a program for a deterministic evaluation of buckling strength and safety. The input parameters are user specified based on the specific track conditions being analyzed. These include lateral resistance, torsional and longitudinal resistances, tie-ballast friction coefficient, track curvature, tie type, rail size, misalignment amplitude and wavelength, rail neutral temperature, maximum expected rail temperature, and vehicle type. Default values are provided for the parameters to guide the user in performing average or baseline calculations. The output includes a buckling curve in the form of a lateral deflection versus temperature increase, which provides information on the track's possible equilibrium (or buckled) states. Pre- and post-buckling longitudinal rail forces and buckling energies are predicted, which are important in some buckling evaluations. The output of CWR-BUCKLE includes two critical temperatures, T_{Bmin} and T_{Bmax} bounding the buckling temperature regimes, allowable or safe rail temperature for buckling prevention, and the buckling margin of safety (i.e., the reserve strength against buckling). Recommendations are provided where the margin of safety is inadequate on methods to increase it.

CWR-INDY is a program for a quick evaluation of CWR track buckling strength and intended for an easier on-the-spot industry use. The inputs for this program are simpler than those required for CWR-BUCKLE. The CWR-INDY inputs are based on information, such as the ballast depth, shoulder width, tie types, track class, and traffic tonnage, which can be more easily obtained and does not require complex measurements. The program converts these inputs internally into their scientific counterparts and uses the CWR-BUCKLE's computational engine to perform the calculations and to evaluate buckling strength and safety. The conversions from CWR-INDY inputs into their scientific CWR-BUCKLE counterparts are based on empirical equations developed through extensive test programs on revenue service lines and tracks at the U.S. Department of Transportation's TTC. The outputs of CWR-INDY are critical temperatures, allowable rail temperature, and a track's (numerical) lateral resistance value. Thus, CWR-INDY also serves as a track lateral resistance estimator, without an actual physical test conducted.

CWR-RISK is a computer program that evaluates the probability of buckling on the basis of statistical principles and input parameters. Since lateral resistance, lateral misalignment, and neutral temperature have been identified as the primary parameters with significant influence on track buckling and which tend to be highly variable quantities, statistical distributions are used to describe them. With these statistical inputs and other deterministic parameters, CWR-RISK evaluates the probability of buckling at a given rail temperature. The analysis identifies a critical temperature, T_c, above which a finite probability of buckling exists.

Probabilistic Approach

The probabilistic approach of CWR-RISK is extended to evaluate the expected number of buckles per year on a given territory. This helps to relate the probability of buckling as a percent occurrence (which is hard to interpret in industry applications) to a more usable parameter, such as the number of expected buckles per year on a specified line segment. This evaluation requires annual temperature and lateral resistance frequencies, misalignment amplitudes, and neutral temperatures in the territory. The application of CWR-RISK shows that the expected annual number of buckles increases with the increase in rail temperature beyond the critical limit T_c. A limiting temperature, T_L, can be chosen by the industry to limit the annual number of buckles to an accepted value.

The probabilistic approach can be used in a decisionmaking process on the imposition of slow orders (i.e., restrictions on vehicle speed during periods of elevated rail temperatures). Presented is a method for selecting the rail temperature up to which full speed may be permissible. Beyond this temperature, a speed reduction regime is specified in which the operating speed is reduced to minimize the possible consequences of a buckle.

Parametric Analyses of Track Buckling

The parametric study, using CWR-BUCKLE, revealed information of practical significance by identifying key variables and evaluating their sensitivity on buckling strength. For example, the peak value of lateral resistance has a significant influence on the upper critical temperature, T_{Bmax}, as has the track's lateral alignment condition. The lower buckling temperature, T_{Bmin}, is influenced by the nonlinear part of the lateral resistance (referred to as the limiting resistance), torsional resistance, and lateral misalignment. Track curvature strongly influences both temperatures. Vehicle truck center spacing and axle loads also impact buckling temperatures through their influence on the central uplift. Study results indicate that for CWR buckling safety considerations, the T_{Bmin} values (hence parameters influencing it) are the more important ones because they drive the safety criterion. Special cases addressing conditions and features such as static versus dynamic buckling, buckling comparisons for weak versus strong lateral resistance track, and wood versus concrete tie track have also been presented with important implications on their respective buckling behavior and the key parameters influencing it.

Safety Limits

Safety limits for buckling prevention using deterministic (CWR-BUCKLE) and probabilistic (CWR-RISK) approaches have been proposed. The deterministic limits are reasonably well developed, whereas the probabilistic approach limits need further evaluation because of a lack of data on parameter variation for the required statistical representations. Deterministic safety limits are based on T_{Bmin} for explosive buckling and T_P for progressive scenarios. The allowable rail temperature, T_{all} (sometimes referred to as the safe rail temperature), is chosen to have a margin of safety of at least 10 °F. The report provides typical safety limit charts to illustrate the concept and the development process.

The safety limits concepts were evaluated and validated by testing various CWR track segments after heating the rails artificially up to the allowable temperature. Two tangent tracks, two 5° curved tracks and one 7.5° curve, were used in the validations, and the results are briefly summarized in the report.

Safety Implementation

To facilitate industry use and application, the safety limits on the allowable rail temperature are translated into limits for the lateral resistance and neutral temperature. For ease of implementation, the requirements on lateral resistance and neutral temperature are uncoupled because the lateral resistance parameter must be above certain minimum values for buckling prevention at a prescribed rail temperature. Based on this approach, this report presents methodologies for monitoring the lateral resistance and neutral temperature.

To monitor the lateral resistance condition, for example, direct measurements on weak or substandard locations are recommended. Under some conditions (such as for recently maintained and stabilized tracks), monitoring can also be accomplished using CWR-INDY's lateral resistance predictor, thereby requiring no physical testing.

Monitoring rail neutral temperature is more difficult because of the lack of an accurate, nondestructive measurement technique. In the interim, the application of existing measurement technologies based on Rail Uplift Device (RUD) in conjunction with strain gage is recommended. The RUD concept, which was field demonstrated in the late 1980s, can be used to calibrate the strain gage without cutting the rail (which is required to obtain the zero calibration value). In situations where the rails are being cut as part of the maintenance action (such as when inserting rail plugs when performing rail repair or destressing), the applications of strain gages is recommended for monitoring the neutral temperatures at those locations. In the event of no measurement being feasible, the assumption of a "neutral temperature safety factor" is recommended for CWR buckling safety evaluation and assurance.

In addition to monitoring methods, this report discusses control of the parameters. For example, the lateral resistance can be controlled to a minimum desired value through several means, including consolidation either through traffic/tonnage or through dynamic track stabilization, using good quality ballast with full ballast section including wide shoulders, and through limiting seasonal curve lateral movement. Better control on neutral temperature variation could be attained through more effective rail break repair and destressing procedures, as well as through minimizing rail movement by providing and maintaining effective longitudinal restraint.

The report also presents conclusions of practical interest and recommendations for further research. Appendices provide additional information on buckling theory formulation/equations and on key parameter measurement techniques and measurement requirements.

1. Introduction

Over the past two and a half decades, significant research work was performed in the field of CWR safety and performance because of its increasing use as a replacement for jointed track. Improved ride quality, increased rail and rolling stock fatigue life, and reduced track maintenance costs were the principal motivations to changing to CWR. However, with the elimination of the joints, the constrained thermal expansion and contraction-induced compressive and tensile forces created the possibility of track buckling in the summer and tensile pull-apart failures in the winter often causing catastrophic derailments. For this reason, substantial research has been conducted to address the CWR safety and performance through the development of analytical models, test investigations, parameter characterization studies, measurement technique development, and improvements in track maintenance practices.

The research work on CWR, sponsored by FRA, has been published in several technical papers and reports. Computer models developed for the buckling safety analyses have been assembled into a program called CWR-SAFE. A user's guide available in the program's HELP menu describes the model's application details. Much of the important technical supportive material has been presented in different sources and therefore must be synthesized into a single comprehensive report to provide an easy updated reference for the research community and to facilitate the understanding and use of CWR-SAFE for buckling safety assessments. This report will fulfill this specific need and further address the most recent developments in probabilistic approaches for estimating track buckling potential, which can be applied for risk-based estimates on buckling safety.

Briefly described in this section are safety-related issues. Present technical problems of CWR review the problem severity from the accident rate perspective and summarize historical background of research work performed to date under the FRA sponsorship.

1.1 Technical Problems

The increased use of the CWR prompted significant research for understanding and controlling failure modes caused by thermal loads in the rails. Two of these failure modes, with a great influence on the safety of vehicle operation, include the following:

- Loss of lateral stability (track buckling, track shift, and radial breathing)
- Rail pull apart (rail break under high tensile forces)

1.1.1 Track Buckling

Use of CWR will generate large thermal rail loads that are typically compressive when the rail temperatures are high and can induce track buckling. The amplitude of the buckling deflections can be large (reaching 30 inches (in) in some cases), and the associated buckle lengths can be on the order of 40–60 feet (ft). These misalignments can cause vehicle derailments because they cannot be negotiated at typical operating speeds.

Track buckling can also occur in the vertical plane, although this mode is rare because of high rail stiffness in the vertical direction. The weight of the track also resists the upward buckling movement. Invariably, whenever the CWR track buckles, it does so in the lateral plane as in Figure 1-1. Even if vertical buckling were to occur, the track would tend to eventually move in the lateral plane because of lower rail stiffness in this plane. The shape of the initial misalignments (lateral alignment defects) is instrumental in determining the resulting buckling mode shape. Rails outside the buckling zone (referred to as the adjoining zones) move longitudinally only with no lateral deflection in these zones. From these adjoining zones, the rails extend and pull into the buckled zone providing the extra rail required for the buckle to take place. The adjoining zones can be as long as 600–1000 ft on either side of the buckled zone, so in a buckling situation, the impacted track length for repair and readjustment can be quite large.

(a)   (b)

(c)   (d)

Figure 1-1. Lateral Track Buckling

A critical issue in the subject of CWR track safety is the influence of vehicle loads on track stability. However, thermal loads alone can precipitate a buckling called static buckling (Figure 1-1(b) and (d)). Buckling due to the combination of thermal and vehicle loads is called dynamic or vehicle-induced buckling (Figure 1-1(a) and (c)). The majority of buckles in revenue service occur under the train; hence, vehicle loads play an important role in the buckling mechanism. This report will present this dynamic buckling theory and its implications on buckling prevention of CWR tracks.

7

In addition to large deflection buckling, track lateral shift and radial breathing of curves can also develop and present problems in CWR tracks.

1.1.2 Track Shift

Apart from thermal and vehicle loads and other parameters to be discussed later, an important contributor to the buckling phenomenon is the initial lateral misalignment. In the case of track buckling, the lateral misalignment influences the buckling strength significantly. The misalignments can have a variety of shapes and can be characterized by a wavelength and its amplitude (offset). FRA has specifications for allowable limits of the misalignment amplitudes. However, these misalignments can trigger buckling especially when they grow. The growth occurs because of repeated application of wheel loads (especially high axle lateral forces), and this growth is usually acerbated by the presence of thermal loads in the rails. The formation and growth of misalignments in CWR is defined as track shift. Track shift can occur and reach a stable limit without leading into full-blown buckling. Conversely, sudden buckling can occur without any visible evidence of track shift. A principal distinction between track shift and track buckling is that track shift safety typically requires the determination of allowable net axle lateral loads (NAL), whereas track buckling safety requires the determination of the allowable thermal loads. Chapter 8 of this report references reports on track shift research performed under FRA sponsorship. The remainder of this report will be confined to the evaluation of CWR track buckling potential and the methods to minimize it.

1.1.3 Radial Breathing of Curves

Curved CWR tracks can experience radial movements or breathing because thermal loads. Compressive loads in the rails generated in summer tend to move the track outward. Likewise, tensile loads in winter will pull the curved track inward. These movements can be large for high degree curves, especially in the presence of high thermal forces and weak lateral restraint. Because radial breathing is seldom uniform, additional misalignments can develop in the track, as well as pockets of lowered lateral resistance. As discussed later, radial breathing is instrumental in the change of the rail's stress-free (or neutral) temperature. As far as the buckling mechanics are concerned, these radial breathing displacements must be accounted for in the proper formulation of the theory for curved track lateral buckling.

1.2 Rail Pull Apart

CWR will not only experience compressive loads because of temperature increase but also tensile forces due to the rail temperature decrease in winter. The maximum tensile load in the rail is determined by the difference in the installation temperature (actually the neutral temperature as defined later) and the lowest rail temperatures. The maximum tensile force can be of the same order of magnitude, or larger, than the maximum compressive load generated in summer. High tensile force is instrumental in causing rail fractures at locations of internal defect and or weak welds. This mode of failure is referred to as pull apart failures. Because of vehicle and thermal stress cycles, the internal defects can grow into sizeable cracks, which propagate rapidly under high tensile load conditions.

The resulting rail gap due to pull apart failures is generally small (usually less than 4 in if the rail is properly anchored or held longitudinally to the ties by effective fasteners). The train wheels may traverse these breaks more safely than they can the large buckled displacements. From the consideration of safety (i.e., derailment potential and consequence), pull apart are considered less severe than buckles. Also, rail breaks may be detected by the signaling system. Hence, the installation temperature of CWR is generally high and sometimes biased to produce smaller compressive loads in summer at the expense of higher tensile loads in winter. Pull apart failures are damaging to track structure and need expensive repairs; therefore, they must be avoided by careful design and maintenance of CWR tracks. The repair of pull apart, especially in cold temperatures, can lead to eventual buckling prone conditions when not properly restressed or adjusted to the correct desired neutral temperature. The industry usually inspects rails for defects by ultrasonic or other means and makes timely replacements/repairs to avoid pull apart and other fracture failures. The tensile pull apart failure prediction is a subject of future research and will not be treated here.

1.3   Track Buckling Accident Severity

The severity of track buckling can be deduced from the statistics of buckling-induced derailments and the resulting dollar damage [1] as Figure 1-2 shows. The data is based on FRA reportable accidents and does not include buckling incidents that did not result in a train derailment. The dollar damage includes the following:

- Loss and damage to the track and the cost of repair
- Loss and damage to the rolling stock and the freight

The damage does not include the loss because of suspended or rerouted traffic.

As can be seen from the data, the past 10 year's average was 34 derailments a year with a decreasing trend over the past 3 years. However, the dollar damage has been increasing over the past 10 years, reaching a peak of $17 million in 2002. The reasons for this trend might be due to better cause-finding and reporting, inflation, and more costly derailments resulting from higher speeds and larger tonnage levels.

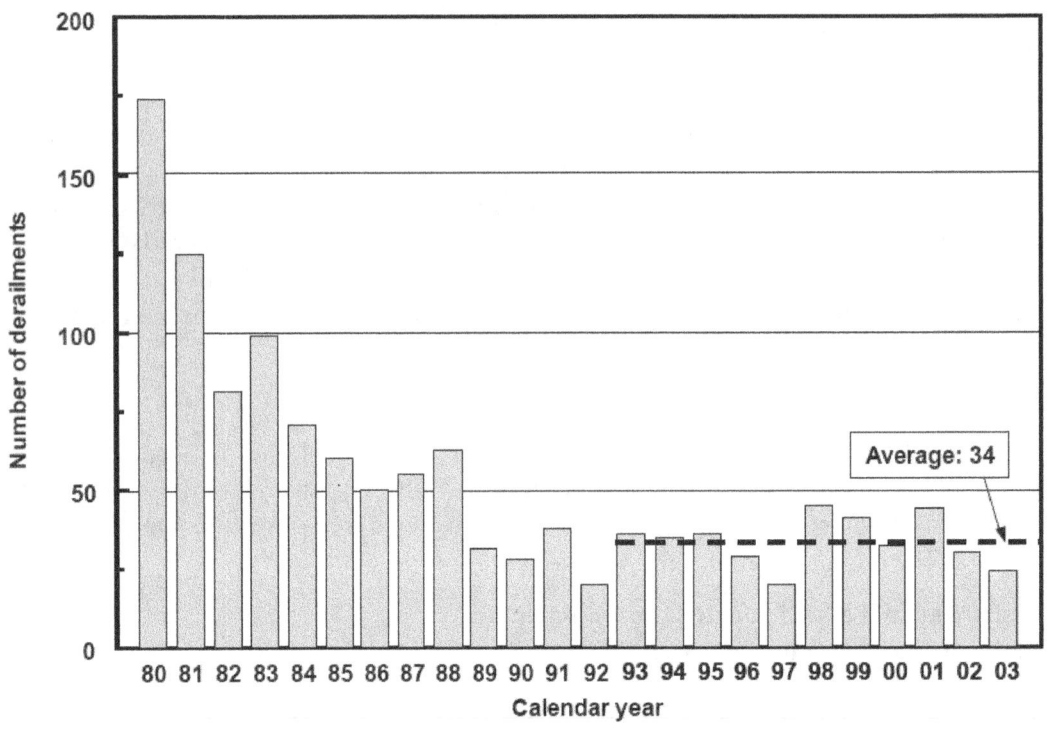

(a) Frequency of buckled track derailments

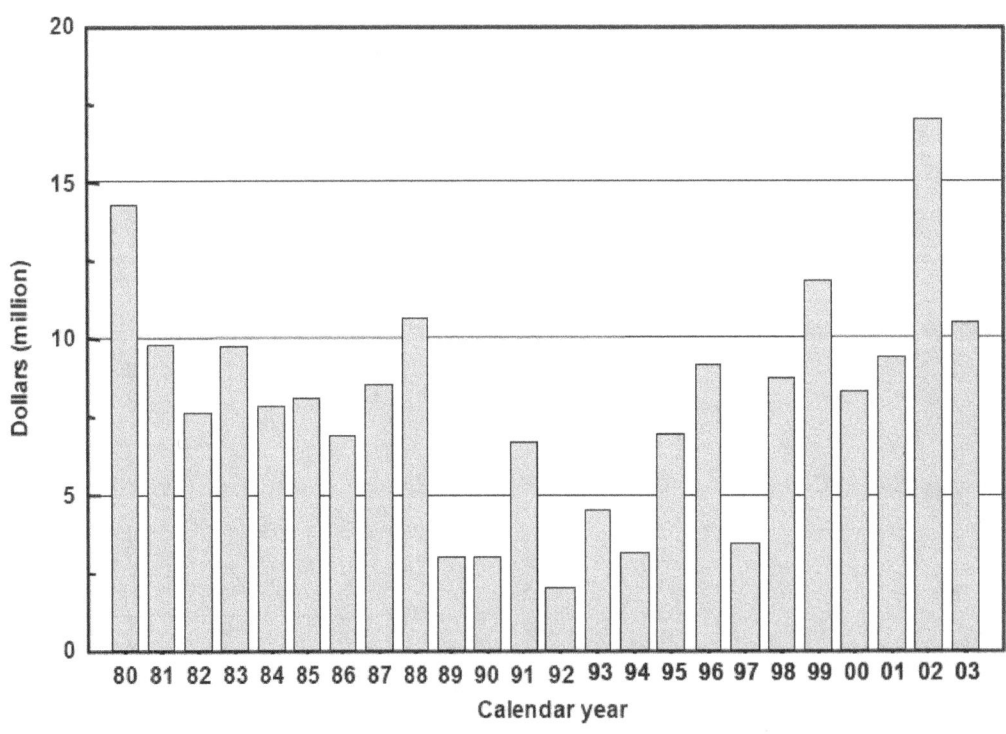

(b) Damage from track buckle derailments

Figure 1-2. Track Buckling Statistics

To keep the derailments and the dollar damage down, FRA and the industry continue to work toward improved buckling prevention practices. This includes a better understanding of the track buckling mechanics and parameters, developing better techniques for improved maintenance, detecting incipient buckling prone conditions, and better managing the risks associated with buckled track. The research presented in this report is aimed at furthering this knowledge toward a better understanding of track buckling prevention.

1.4   Background

Under FRA's sponsorship, significant research and development work has been performed in the area of CWR buckling. The work involved development of the theory for predicting the critical forces and temperatures causing buckling, identification of critical parameters in the theory, development of a safety criterion, conducting full-scale static and dynamic buckling tests for model and safety criterion validation, and development of methodologies for the assurance of safety against buckling. A comprehensive computer tool (CWR-SAFE) is a major achievement under this FRA program, as are new techniques and hardware for measuring key parameters required in the computer program and methods for implementing improvements in the maintenance practices by the industry.

1.5   CWR Theory

The early theories of buckling, which do not include vehicle load effects, are known as static theories. The early version of static theory was developed in reference [2] for tangent and curved track with misalignments. The theory was validated with tests at Plains, VA, in 1982 on the Southern Railroad [3]. Because track buckling is predominately a train-induced event, the vehicle loads were considered to be important. This resulted in a dynamic theory of CWR track buckling by Kish et al. in 1984 [4]. Several tests on tangent and curved tracks were conducted [5–7] at TTC in Pueblo, CO, with the Association of American Railroads (AAR)/TTC test support to validate the dynamic buckling theory. New safety concepts were also developed based on these works. Reference [8] summarizes the dynamic buckling theory, test validations, and safety concepts. To apply the theory, several computer programs were developed, culminating in CWR-SAFE.

Using the dynamic buckling theory, an extensive study was conducted to identify critical parameters and evaluate the sensitivity of the parameters on buckling strength [9]. The following section briefly reviews theses parameters, and a later chapter presents a comprehensive discussion on them.

1.6   Buckling Parameters

Extensive field tests were conducted to evaluate and characterize track parameters having large influences on buckling strength. These included the following:

- Tie-ballast lateral resistance
- Rail fastener longitudinal and torsional resistance
- CWR neutral temperature

New specialized, portable hardware was designed and developed for field evaluation of single tie lateral resistance for wood and concrete ties. The tie-ballast resistance data was developed for revenue service conditions and TTC track. The data encompasses a variety of track conditions and parameters, such as shoulder width, different crib levels, pre- and post-tamping, different consolidation levels, and different ballast materials. The results and the description of the hardware are available in references [10, 11].

The longitudinal and torsional resistance of rail fasteners and ties in the ballast was evaluated at TTC, under laboratory and field conditions. The resistances were measured for various wood and concrete fasteners and conditions [9].

Using rail force measuring strain gages, the CWR neutral temperature was monitored on several revenue lines and at TTC for tangent and high degree curves. Conditions under which the neutral temperature can significantly drop from the installation value were evaluated [12]. These measurements were instrumental in providing information on the causes of neutral temperature variation and on typical values. Subsequent chapters in this report will address further discussion on these and other parameters, including their influences on track buckling.

1.7   Computer Programs

CWR-SAFE is a PC-based analysis program that calculates the buckling response of CWR tangent and curved track because of thermal and vehicle loads. The software embodies three analysis modules called CWR-BUCKLE, CWR-INDY, and CWR-RISK. CWR-BUCKLE performs buckling analysis, determines the allowable temperature increase for buckling prevention, performs safety analysis to determine buckling safety margin (BSM), and serves as a platform for the CWR-INDY and CWR-RISK modules. Embedded in the software is a track quality-based safety criterion, which is used to determine the allowable temperature increase or the safe temperatures for buckling prevention.

CWR-INDY is a simpler version of CWR-BUCKLE, customized for easy industry use by allowing for simple design parameter inputs such as:

- Tie, fastener, and ballast type
- Crib and shoulder width
- Consolidation type and level

These input parameters are converted to their scientific numeric counterparts within the program using empirical equations based on field test data. With the converted scientific data, CWR-INDY predicts the same type buckling estimates as CWR-BUCKLE.

Although CWR-BUCKLE and CWR-INDY are deterministic models, the CWR-RISK module provides a probabilistic analysis of track buckling. Statistical descriptors are used to define the three important parameters lateral resistance, neutral temperature, and lateral alignment defect. With these input parameters, CWR-RISK predicts the probability of buckling as a function of the rail temperature and allows for the evaluation of various risk acceptance levels to be used in buckling prevention. These risk levels can then be used for the development of risk-based performance standards for CWR or the determination of risk-based slow orders for hot weather operation [13].

1.8   CWR Maintenance

In addition to the development of theoretical models, hardware, and applications programs for field measurement of parameters, the research has also produced important advances on determination of longitudinal rail force and its influence on longitudinal restraint of CWR, aimed at helping improve CWR maintenance and inspection practices.

1.8.1 Rail Longitudinal Force Evaluation

A significant aspect of buckling prevention is the effective control of the longitudinal force in CWR to safe levels. The allowable or safe longitudinal force level can be determined through CWR-SAFE, and these then can be monitored through accepted measurement diagnostics.

Under FRA's research program, the Rail Uplift Technique was developed to measure rail force [14]. The technique is based on relating the CWR tensile or compressive force in the rail to the uplift deflection of the rail when freed from fasteners and pulled up by a car-mounted uplift device. The method was successfully demonstrated at TTC and was used on the Burlington Northern Railroad [15]. It is the only known method to give the absolute force of the rail without cutting the rail. However, because the fasteners needed to be removed to facilitate the rail uplift, the method received minimal acceptance by the U.S. railroad community.

Techniques, based on strain gage concepts, have been used to measure rail force, although with difficulty in application because of the need to have a zero reference, which requires cutting the rail. Other techniques explored based on ultrasonics, acoustics, magneto-elasto-mechanics, Barkhausen-noise, vibro-elastics, x-ray diffraction, and fiber optics were proven ineffective, except in some specific laboratory applications. Hence, to date, a nondestructive, accurate, and easily deployable longitudinal force measurement system, either vehicle or track borne, is not available, making the development of CWR force measurement a major worldwide research need.

1.8.2 Destressing of CWR

One way railroads attempt to control large compressive stresses in the summer is to cut, destress, and reweld the rail after adjusting it to the desired level of neutral temperature. In winter conditions, large tensile forces may fracture the rail, requiring repair and readjusting to the desired neutral temperature. Under FRA's research program, a longitudinal restraint model has been developed for a more effective destressing of rails in summer and restressing in winter [16]. A prototype for a portable device called the Rail Destressing Force Indicator (RDFI) has been designed to help the track crew to obtain the required information for a more effective rail destress/readjustment procedure. An RDFI prototype has been field tested and is currently being finalized for industry trial.

1.9   Report Content and Organization

Chapter 2 presents the theory of track stability that is required in buckling safety analyses, which forms the basis of the CWR-SAFE buckling model. The theory considers the vehicle and thermal loads on tangent or curved CWR with lateral misalignments and nonlinear lateral resistance characteristics. Appendix A presents the actual mathematical formulations, but this section presents the mechanisms and assumptions in the development of the theory. This section also presents the procedure used to determine the buckling strength (in terms of critical forces and temperatures), along with a fundamental safety criterion for buckling prevention and the key parameters influencing the criterion.

Chapter 3 presents an overview of CWR-SAFE and its three modules. The model's applications are illustrated with numerical examples. Chapter 3 discusses assumptions and correlations of the CWR-BUCKLE scientific parameters to simpler design counterparts (such as the ballast and fastener type, crib content, and shoulder width), which are the required inputs to CWR-INDY. This chapter also briefly introduces the CWR-RISK fundamentals.

Chapter 4 presents detailed parametric and sensitivity studies on the basis of the CWR-BUCKLE program. Chapter 4 describes the effects of track resistance, misalignments, curvature, and vehicle parameters on track buckling strength. Appendix B presents the techniques to measure the parameters.

Chapter 5 presents the CWR-RISK module that deals with the probabilistic buckling theory. Chapter 5 describes the assumptions, the details of the methodology, and the probabilistic descriptors for the parameters, including implications on buckling safety. As an example of the model's application, a risk-based approach for the slow orders is also provided. Appendix C presents the mathematical formulations involved in CWR-RISK. Appendix D presents the measurement requirements including sampling criteria for lateral resistance and neutral temperature parameters required in CWR-RISK.

Chapter 6 deals with the safety aspects of CWR. The chapter presents applications of the buckling safety assurance methodology, including buckling prevention safety limits, as well as other CWR safety applications, such as track consolidation, hot weather slow orders, and neutral temperature maintenance.

Chapter 7 details the conclusions on all aspects of CWR studied to date in the United States. This chapter also provides recommendations and future research needs.

Chapter 8 lists all the references for this document.

Finally, Appendix E addresses neutral temperature issues during CWR maintenance operations, such as rail repair via cutting, destressing, and readjusting. This appendix also discusses the RDFI device for a more effective destressing procedure.

2.   The Theory of Track Stability

This chapter describes a track buckling mechanism, theories for static and dynamic buckling, and governing parameters. The chapter also introduces the buckling energy concept, buckling safety criteria, and the analysis methodology for conducting buckling safety evaluations.

2.1   Buckling Mechanism

Consider a long CWR track, Figure 2-1, which is straight but for a small initial lateral sinusoidal type misalignment described by an amplitude δ_0 and a wavelength $2L_0$. With increase in rail temperature, the compressive force P will increase, which may produce some growth in the initial misalignment. Experiments and field observations have shown that as temperature (and the corresponding rail force) increase to a maximum (critical) level will increase the initial misalignments to w_B, an unstable equilibrium state. At this state, the track can buckle out suddenly into a new lateral position, w_C, spanning a length of 2L. The magnitude of w_C is typically large, on the order of 6–30 in, while its wavelength can be on the order of 40–80 ft. This process, in terms of temperature increase above neutral versus the maximum deflection, is graphically shown as in Figure 2-2(a). The neutral temperature (as referred to on the vertical axis) is *that temperature at which the net longitudinal force in that rail is zero* and is a reference condition for the force build up in the rail; as shown later, this is a key parameter for buckling prevention. The displacement w_B is referred to as the pre-buckling displacement, the temperature increase at B when buckling takes place is the buckling temperature, and w_C is the post buckling displacement.

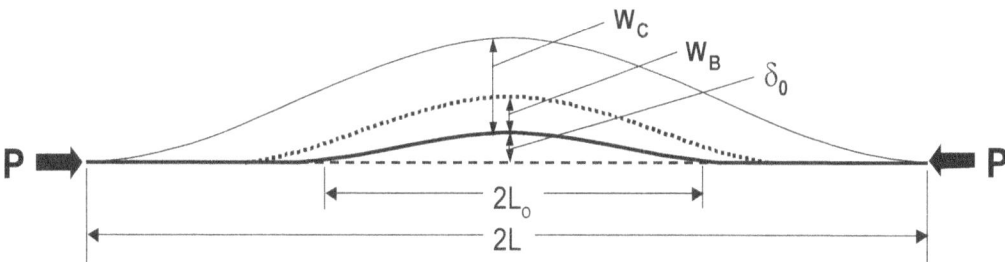

Figure 2-1. Pre- and Postbuckled Track Configurations

Figure 2-2(b) shows the same process with the dashed overlay representing all the possible buckled positions (equilibrium configurations) given by stability theory (to be discussed later). Two distinct temperature increase values ΔT_{Bmax} (sometimes referred to as the bifurcation temperature) and ΔT_{Bmin}, in between which multiple positions of equilibrium can exist. A unique property of buckled or equilibrium positions is the nature of their stability (i.e., being stable or unstable). Thus, the dashed prebuckled branch up ΔT_{Bmax} can be shown as stable, the dotted branch down to ΔT_{Bmin} unstable, and the dashed branch increasing up to C and beyond to be stable. In actual physical conditions, only the stable positions can be realized. The bifurcation point B represents the configuration common to stable and unstable configurations (sometimes referred to as infinitesimal stability) at which the track snaps over to the stable position at C. Research has also shown that, if sufficient external energy is supplied (such as by train action), the track can jump from a prebuckling stable configuration to a postbuckling stable configuration (through unstable configuration) at temperatures below ΔT_{Bmax} as shown 1 to 2 to 3 in Figure 2-

2(c). Hence, the range between ΔT_{Bmin} and ΔT_{Bmax} represents the buckling regime of CWR tracks. Whereas the track will buckle at ΔT_{Bmax} with no external energy, it can also buckle at ΔT_{Bmin} if sufficient external energy is supplied to the track. Below ΔT_{Bmin}, the track will not buckle because it has only one stable equilibrium configuration.

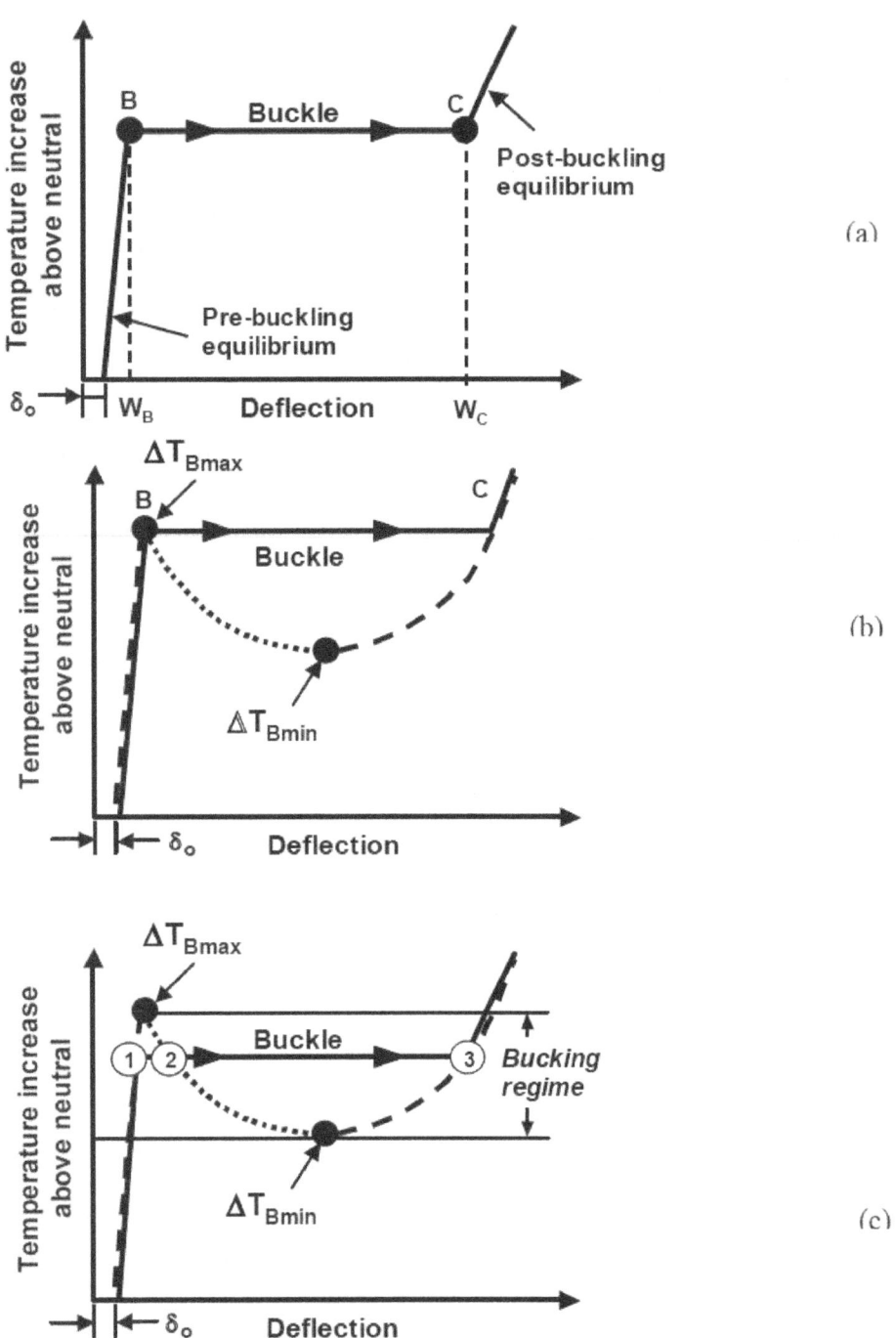

Figure 2-2. Buckling Response Curves

The shape of the buckling response curve depends on the specific track parameters and conditions (i.e., track quality). Good quality track will have a large difference between the maximum buckling temperature (ΔT_{Bmax}), and the minimum buckling temperature (ΔT_{Bmin}) as shown in Figure 2-3 (a). As track quality decreases, both the ΔT_{Bmax} and the ΔT_{Bmin} value decrease as does the Δ between them. In fact, for low or inferior quality track, the difference between these temperatures can go to zero (when ΔT_{Bmax} and the ΔT_{Bmin} coalesce into one value ΔT_P), as shown in Figure 2-3(c). Such a track will buckle out slowly or progressively with the rail temperature rise (i.e., no sudden snap through).

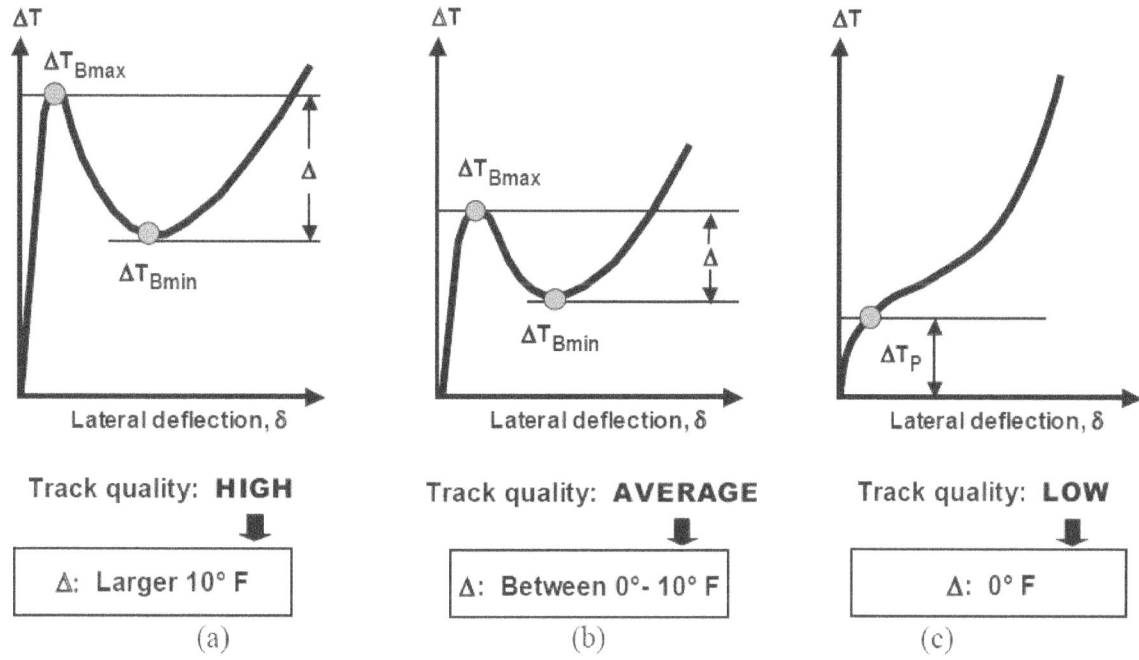

Figure 2-3. Track Quality Influence on Buckling Response Characteristics

2.2 Force Drop/Buckle Influence Zone

A feature of the sudden explosive buckles is the accompanying rail force drop (energy release) in the buckled zone compared with that of the prebuckling force value. This is due to the large lateral displacement contributing to the rail extension that releases some of the compressive load. The lateral displacement in the buckling zone is accompanied by the longitudinal motion in the outside zones, which feeds the rail into the buckling zone. The longitudinal motion will be felt through a substantially long section of CWR track. Thus, the rail force distribution in the buckled and adjoining zones is significantly altered, as indicated in Figure 2-4. This means that the CWR neutral temperature will be significantly altered after the buckling incident, and long sections of track have to be repaired and restressed. It also means that the theory of track buckling has to appropriately predict this energy release and force drop to correctly model buckling mechanism.

17

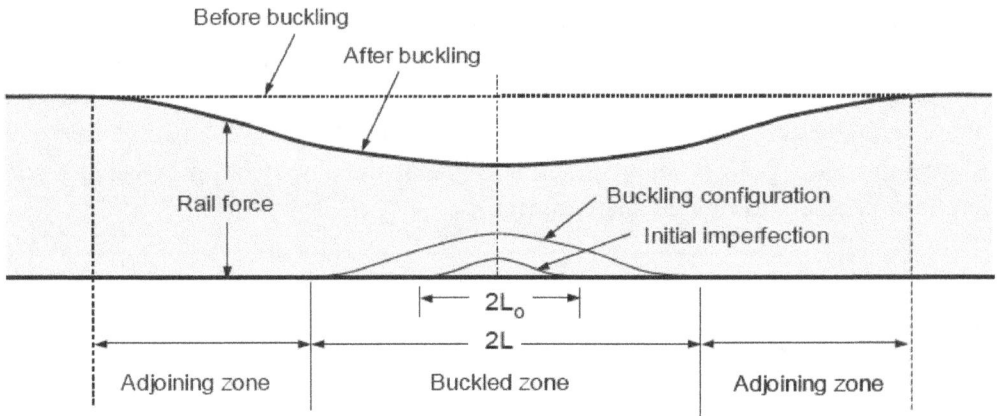

Figure 2-4. Rail Force Distribution after Buckling

2.3 Buckling Mode Shapes

Figure 2-1 depicts the mode shape as a symmetric half sine wave called Shape I. The mode shape can also be a complete sine wave reflecting an asymmetric wave called Shape II, whereas higher modes, such as symmetric Shape III, can also occur. Figure 2-5 shows these schematically. The actual buckled mode shape occurring in track is largely influenced by the shape of the initial misalignment. Tangent track typically buckles out in Shape III, whereas curved track buckles in Shape I, and the theory has to properly account for them in terms of appropriate boundary conditions, as discussed later.

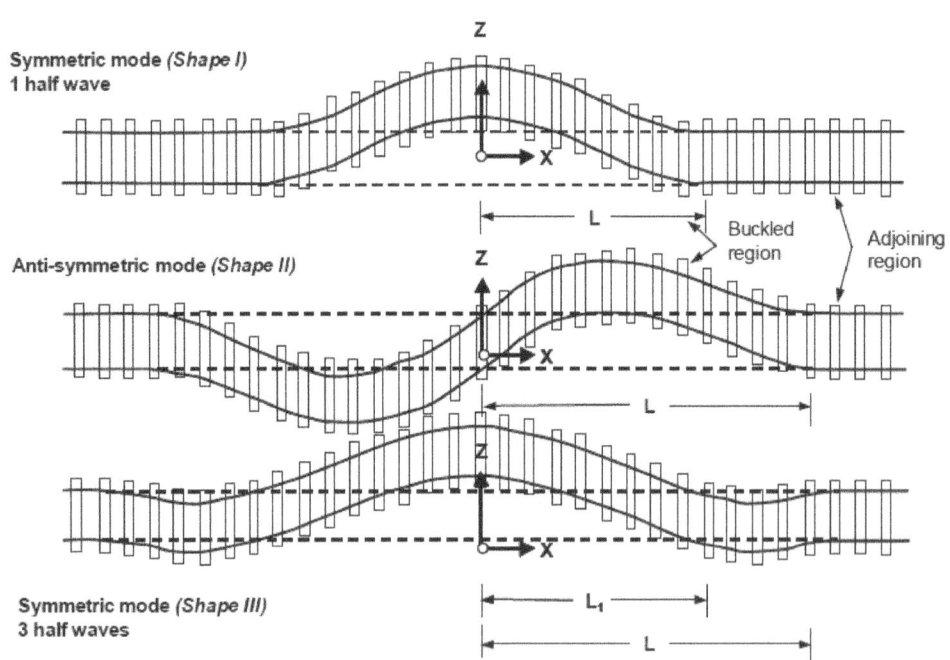

Figure 2-5. CWR Track Lateral Buckling Modes

18

2.4 Track Buckling Theories

Buckling theories are based on mechanistic track model shown in Figure 2-6. The buckling force is the combined compressive load in the two rails, which depends on the rail cross-sectional area and the temperature rise. The lateral resistance generated between the ties and the ballast, as well as the longitudinal and torsional resistances generated in the rail fasteners, offers the resistive forces to the buckling force.

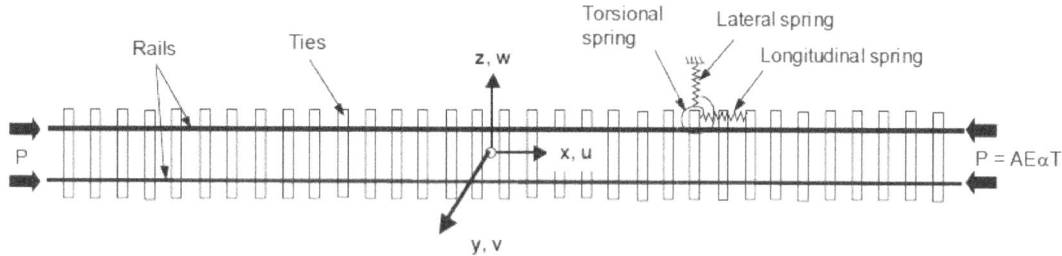

Figure 2-6. Track Model

The lateral resistance depends on nonlinear tie-ballast spring characteristic because the tie displaces laterally through the ballast. The longitudinal resistance depends on longitudinal spring characteristic of the rail/tie/fastener/ballast. In a poorly ballasted condition, such as with low ballast level in the cribs, the rail-tie structure could move longitudinally. In other cases, the rail creeps through the fasteners with no or negligible longitudinal movement of the ties. The rail-to-tie fastenings also offer rotational rigidity (modeled by torsional springs), which reacts against the rail's tendency to rotate during the buckling deformations.

A number of theories have been postulated in the literature on track buckling. A review of all the theories to date is available in reference [17]. The theories developed under FRA research are divided into two basic categories, deterministic and probabilistic.

The deterministic method requires parameters to be specified with certainty as in classical mechanics. The buckling strength of the CWR track is evaluated using classical formulations and expressed in terms of the temperature increase over the neutral temperature (i.e., the buckling response curves of Figure 2-3). The method indicates that either track will buckle or not at a given temperature for a set of input parameters. A later subsection presents the required parameters and their characterization.

The probabilistic method provides a percent probability of occurrence of buckling at a given rail temperature but requires a statistical distribution of input parameters. The probabilistic theory can thus facilitate a risk-based approach to vehicle operations and maintenance procedures to manage the buckling potential. The deterministic and probabilistic methods can be used for static and dynamic buckling analyses and for tangent and curved tracks, as discussed in subsequent sections.

2.5 Static versus Dynamic Buckling Model

The static buckling model ignores the effects of vehicle loads and considers the buckling of track resulting from longitudinal compressive loads only. In the buckling region, the resistance to lateral buckling is offered by a tie-ballast structure with no vehicle vertical load influence. Although this mode of buckling has been extensively treated in the literature and in test studies, it is incorrect to use these results to explain buckling occurring in conjunction with train movements. The correct modeling requires the inclusion of the vehicle vertical loads and their influence on lateral resistance. The vertical loads are distributed along the track depending on the axle and truck center spacing of the vehicle. Although the lateral resistance directly under the wheels increases, the resistance of the track segment between the trucks can decrease because of the uplift wave of the section. This uplift wave is referred to as the central wave and shown schematically in Figure 2-7. Recession and precession waves exist (or occur) behind and in front of a train (as shown for two cars in the figure) where the lateral resistance can reduce. The central wave generally has maximum reduction in the lateral resistance and is therefore critical for buckling. This suggests that the track section under the vehicle can be most vulnerable to buckling due to loss of lateral resistance; in fact, buckling under the train is a frequent occurrence.

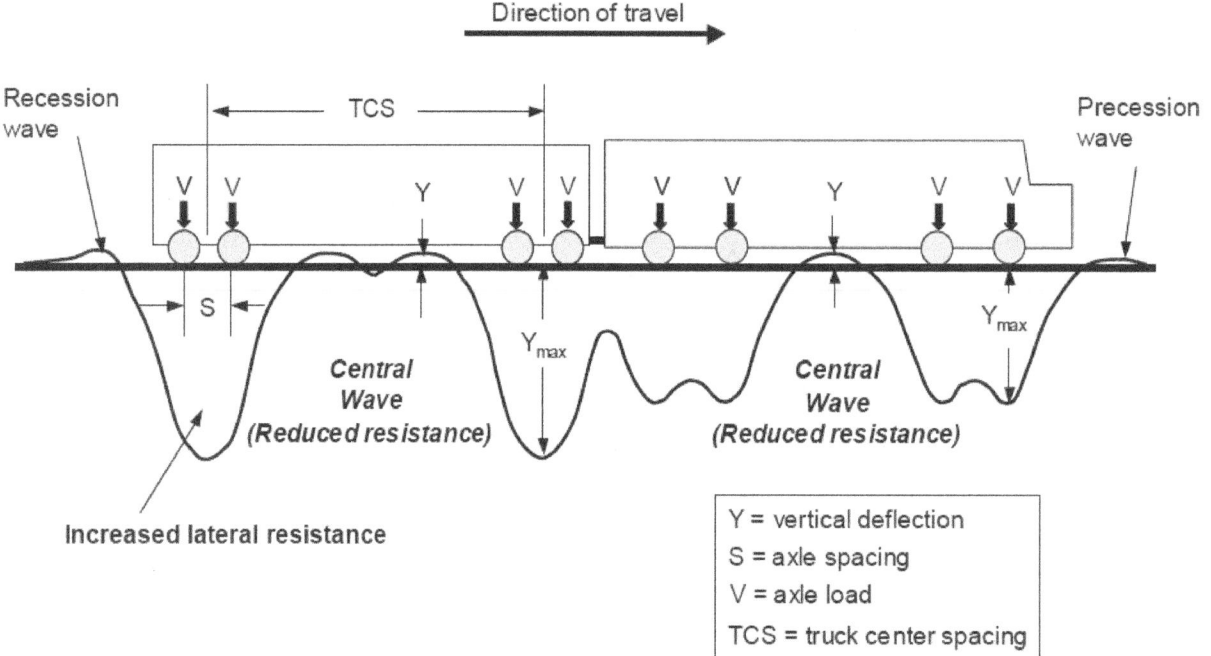

Figure 2-7. Definition of Uplift Waves

Buckling under the central wave uplift should be distinguished from the potential track shift that could occur if the wheels carry high lateral loads during negotiation of curves and lateral misalignments. The force that tends to move the curve laterally is the net axle force (the sum of the lateral forces at the two wheels of an axle). The net axle force is expressed in terms of the lateral to vertical force ratio (L/V). Unless this L/V ratio exceeds a threshold limit, no track shift or lateral movement will occur under the wheels [18]. Track shift occurs gradually under a wheel passage, whereas lateral buckling between the two trucks of a car can occur rather suddenly. Track buckling due to central wave uplift is generally not dependent on the NAL force because the lateral deflection, if any, from the NAL load is confined to a small region under the wheel where the lateral resistance is the highest.

In summary, dynamic buckling theory is required for the more accurate buckling predictions. This means that the influence of the vehicle loads in producing the dynamic uplift is essential in the model, specifically, to account for the reduced resistance and the subsequent reduction in buckling strength.

2.6 Tangent versus Curved Tracks

Field tests and observations of actual buckles show that the buckling behavior of curved tracks can be different from that of tangent tracks. Tangent track is generally a sudden explosive type of buckle and can displace to either side depending on the direction of the initial misalignments or the weaker side of lateral resistance. Track buckles with Shape III on tangent track typically have the amplitude of the middle wave relatively large compared with the two end waves.

Curved tracks generally buckle outward in Shape I. Because of the initial curvature, it would require significant energy to bend the rail in the opposite direction of its curvature (i.e., inward to produce the tail ends of Shape III). Another important feature of curved track buckling is the tendency toward progressive buckling, especially for curves with weak lateral resistance and with high curvatures.

Curved tracks also exhibit radial movement (breathing), especially under weak lateral resistance condition when in the presence of large diurnal and seasonal rail temperature changes. Temperature increase over the neutral can produce radially outward movement; temperature decrease from the neutral can produce radially inward movement. This radial breathing can be detrimental to the track because it reduces the lateral resistance further and may generate local lateral misalignments, which can precipitate buckling. Radial movement can be on the order of a few inches or more and can create buckling prone conditions by reducing neutral temperature, weakening lateral resistance, and producing local alignment defects.

2.7 Differential Equation Formulations

Buckling theories are based on the differential equations of equilibrium presented in Appendix A for tangent and curved tracks. These equations are derived using the variational principles of minimizing the potential energy of the track system. One differential equation is for the transverse or lateral displacement in the buckled zone. The second one is for the longitudinal or tangential displacement in the adjoining zone. These equations are then coupled through another equation for the temperature derived by using continuity requirements between the buckled and adjoining zones. These equations are based on the classical beam theory allowing for large lateral buckling displacements (requiring nonlinear strain-displacement relationship). The following key assumptions are made in the buckling theory:

- The two rails are represented as a beam with a combined moment of inertia and a combined cross-sectional area. The longitudinal force in the beam is the sum of the forces in the two rails. Rail cross sections before buckling remain plane after buckling.
- Lateral resistance offered by the tie is represented by a nonlinear (spring) idealization closely approximating test measurements.
- Longitudinal resistance offered by the fasteners and ballast to the rail can be represented by a linear idealization for the range of longitudinal displacements (0.5 in) typically found in buckling tests.
- Ties do not deform or rotate, and the connection formed by the fasteners to the rails is represented by a torsional spring. The torsional resistance in the tie-rail fastener is proportional to the rail rotation for small angles (<5°) observed in the buckling tests and in tests conducted on fasteners in the laboratory.
- Rail force is constant over the buckling zone length as compared with the adjoining zone, where the force varies in a manner depending on the longitudinal resistance of the fasteners.
- Rail temperature is uniform (longitudinally and across the rails) throughout.

- Adjoining zones in the tangent track experiences longitudinal displacement only with no lateral movement. For curved track, the adjoining zone experiences uniform radial movement in addition to the tangential displacement.
- The track has self-weight and no significant vertical profile irregularities.
- Lateral track misalignment has a known wavelength and amplitude, and the shape is approximated mathematically by a polynomial function.

2.8   Buckling Response Computation

To determine the buckling response, the relationship between the temperature rise and lateral displacement (Figure 2-2b) must be evaluated. For this purpose, different equilibrium configurations with different wavelengths are considered. Each buckling configuration gives a point on Figure 2-2b.

The differential equation for the buckling zone connects the rail force and the lateral displacement. The solution, $w = w(x)$, of the differential equation represents buckling wave shapes. For a specific wave shape such as Shape I, the solution gives a relationship between the wavelength, L, buckling amplitude, w, and the rail force, $\bar{P}$, in the buckled zone. Using continuity conditions on longitudinal displacement (and its gradient) between the buckling zone and the adjoining zones, a temperature equation relating the rail force, $\bar{P}$, in the buckled zone, rail temperature, T, and the wavelength, L, is derived. By varying L over a range, the corresponding values of the buckling force, $\bar{P}$, the amplitude, w, and the rail temperature, T, can be determined. From this data, the buckling response curve connecting the rail temperature and the buckling amplitude can be constructed.

The complexity of determining the response curve depends on the numerical scheme adopted. The most convenient approach is the application of Fourier series for the solution of the differential equations, as shown in Appendix A. The advantages of the series approach including the following:

- It can accommodate any nonlinearity in the lateral resistance and variations of the resistance under the wheel loads.
- The method can predict explosive or progressive response for straight and curved tracks with misalignments.
- The method works for perfectly straight tracks with no initial misalignments.
- The computational time is very small (under a few minutes) and the method facilitates easy programming on a PC.

An alternate approach to the foregoing method is the finite element (FE) method. The nonlinear FE method uses an incremental approach (i.e., the equations are formulated in terms of increment in temperature rise and lateral deflection). Temperature or load increments are selected, and a relationship is determined starting from the initial condition of zero temperature. This approach requires some misalignment to be prescribed a priori and may face numerical difficulties because of the instability at ΔT_{Bmax}. The FE method requires significant computational time, and its accuracy is not generally as good as the Fourier series approach used here.

22

2.9   Parameters Influencing Buckling Response

Foundation Modulus (Vertical)

The track foundation modulus is a measure of the vertical stiffness of the track foundation. This is used to determine the vertical load distribution on the ties, from which the dynamic lateral resistance is evaluated.

Lateral Resistance (Peak and Limiting)

The tendency of the track to buckle laterally is resisted by the reaction forces exerted by the ballast on the ties. The ballast lateral resistance typically exhibits characteristics such as those shown in Figure 2-8. The consolidated track exhibits a distinct droop in resistance after the peak value, whereas freshly tamped track does not, it continues at the peak value. Figure 2-8a also shows the subparameters used to define the lateral resistance characteristic of consolidated track, including the following:

- Peak Lateral Resistance, F_P, and corresponding displacement, w_P
- Limiting Lateral Resistance, F_L, and corresponding displacement, w_L

For weak or recently maintained tracks, tests have shown that $F_P = F_L$ and $w_P = w_L$.

In the measurement of lateral resistance, as discussed later, it is generally easy to determine F_P and w_P by loading the tie laterally by no more than 0.5 in. To determine F_L and w_L, 3–5 in of tie movement may be required, which is considered more destructive to the track. Hence, through extensive test work, correlations were developed to relate F_P, w_L, and F_L (see Chapter 3), so in many cases, F_P is adequate to define the resistance characteristic. Chapter 4 further discusses the importance of F_L.

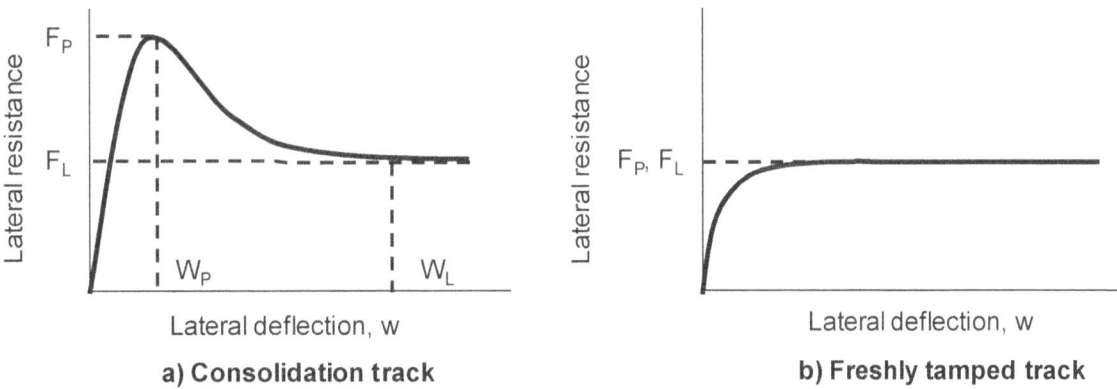

Figure 2-8. Typical Lateral Resistance Characteristics

Longitudinal Resistance

Track longitudinal resistance is the resistance offered by ties and ballast to the rails because they tend to move in the longitudinal direction in the event of buckling, when thermal forces cause longitudinal movements, or in response to braking and accelerating train action. The resistance characteristic is generally bilinear, and in most cases, only the initial linear part will be required because the rail longitudinal displacement in the buckling event is small.

Misalignment Amplitude

The misalignment amplitude (δ_0) is the size of the track misalignment prior to the occurrence of buckling, as shown in Figure 2-9.

Misalignment Wavelength

The misalignment wavelength ($2L_0$) is the total length of the track misalignment before the occurrence of buckling, as shown in Figure 2-9.

The misalignment shape can be mathematically approximated by a simple polynomial, as shown in Appendix A, giving zero values of amplitude and slope at the ends of the misalignment wave.

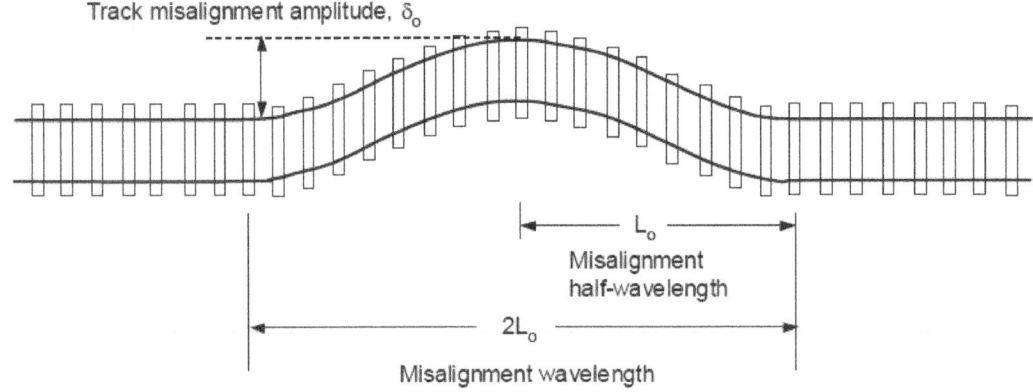

Figure 2-9. Track Misalignment

Neutral Temperature (T_N)

The neutral temperature is the rail temperature at which zero longitudinal force exists in the rail. The temperature increases in the buckling analysis are measured with respect to the neutral temperature. Although this parameter is required only in the safety analysis part of the deterministic model, a statistical distribution of this parameter is a key input in the probabilistic buckling strength evaluations.

Rail Size

Figure 2-10 shows the coordinate axes for a single rail beam and some rail properties (area and moments of inertia).

	Rail	Area (in^2)	Moment of Inertia (in^4) (about the horizontal axis, I_{yy})	Moment of Inertia (in^4) (about the vertical axis, I_{zz})
Y (vertical) X (longitudinal) Z (horizontal)	115 RE	11.25	65.6	10.8
	119 RE	11.65	71.4	10.9
	132 RE	12.95	88.2	14.6
	136 RE	13.35	94.9	14.7
	UIC60	11.91	73.4	12.3

Figure 2-10. Rail Beam Sectional Properties

Rail Temperature (T_R)

The rail temperature is the maximum anticipated temperature of the rail found in the segment of track being analyzed. This is required together with the rail neutral temperature in the safety assessment parts of the analysis.

Tie-Ballast Friction Coefficient

The tie-ballast friction coefficient is a measure of the tie bottom surface influence in lateral resistance when the track is vertically loaded. It is required in the evaluation of the track dynamic lateral resistance and the uplift computation.

Torsional Resistance

The rail fasteners (which can include cut spikes/anchors or elastic fasteners) also provide rotational restraint against the lateral bending of the rails. The resistance is linearized with respect to the rail rotation, and the torsional stiffness is specified per fastener. The required input torsional resistance is pounds per unit length of track and is calculated from

$$\text{torsional resistance} = 2\frac{\text{torsional stiffness}}{\text{tie spacing}}$$

where the factor 2 accounts for both rails.

Track Self-Weight

The track self-weight is the weight per unit length (pounds per inch) of the track structure, including the rails, fasteners, and ties. The following formula is used for its determination

$$\text{self weight} = \frac{W_{TIE}}{S} + \frac{2\left(W_{RAIL}\right)}{36}$$

where

W_{TIE} = weight of one tie including fasteners and tie plates
W_{RAIL} = rail weight (lb/yard)
S = tie spacing (inches)

Vehicle Characteristics

The required vehicle characteristics for dynamic buckling theory are the truck center spacing, the axle spacing, and the axle loads for each car. These determine the uplift characteristic in the central zone where buckling is likely to occur.

2.10   Buckling Safety Criteria

2.10.1 Background

Referring to Figure 2-2b, if the rails are heated to the upper critical temperature, ΔT_{Bmax}, corresponding to the point B, the CWR track will be in a state of unstable equilibrium, meaning that the track will jump to the configuration corresponding to the point C with no external energy supplied to it. If it is heated to a lower point in the buckling regime, as indicated in Figure 2-2c, some finite external energy must be supplied in the lateral plane to buckle the track laterally (to a point on the stable post buckling branch).

Clearly, trains operating at their line speeds supply energy in the lateral plane because of wheel set lateral movement. This vehicle energy can be accentuated if traversing misalignments or when negotiating curves. The energy imparted to the track is generally proportional to the square of the train speed. At higher speeds, higher levels of energy may be imparted to the track producing buckling at temperatures lower than ΔT_{Bmax}, typically under trains. Using this ΔT_{Bmax} temperature as an allowable or safe temperature is inadmissible because of this variability (as well as to other sensitivities as shown later).

The energy required to buckle the track laterally as the rail temperature increases within the buckling regime ($\Delta T_{Bmax} \geq \Delta T_r \geq \Delta T_{Bmin}$) can be computed by using the expressions developed in Appendix A. Figure 2-11 shows the result of such energy computation, showing the energy required to buckle the track for certain assumed parameters as a function of the temperature increase above neutral temperature. This shows zero energy at ΔT_{Bmax} as expected at the point of unstable equilibrium, the maximum energy at ΔT_{Bmin}, and a very sharp decrease in energy required for buckling just above ΔT_{Bmin}.

Buckling energy computations have indicated that the magnitude of the maximum energy (referred to as the energy barrier) at ΔT_{Bmin} increases as the buckling regime ($\Delta T_{Bmax} - \Delta T_{Bmin}$) increases. This leads to the expectation (partially verified by analysis) that, for track conditions exhibiting large buckling regimes, the external (train) energy will not be larger than this maximum, hence not sufficient to buckle that track at ΔT_{Bmin}. However, above ΔT_{Bmin} buckling will be possible due to the rapidly decreasing energy levels that are possible because of train passage.

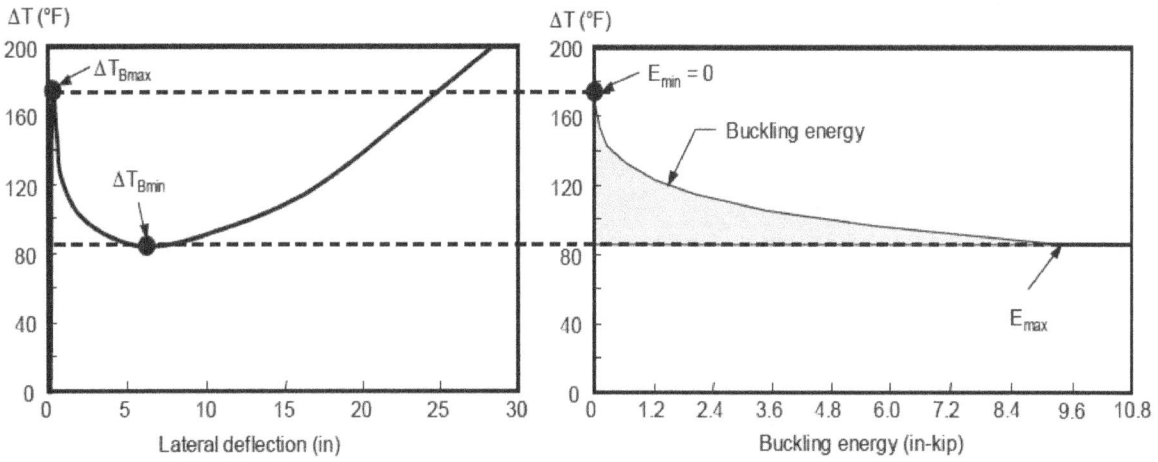

Figure 2-11. Energy Required to Buckle

This aspect is incorporated in the development of the safety criterion presented below, which is predicated on the fact that the safe temperature level for buckling prevention can be based on ΔT_{Bmin}.

2.10.2 Buckling Safety Criteria

A buckling safety criterion is postulated based on the following stipulations:

- The permissible rail temperature rise cannot be greater than the lower buckling temperature ΔT_{Bmin} because of the possibility of buckling at this or higher rail temperature.

- An adequate energy barrier should exist at ΔT_{Bmin} if ΔT_{Bmin} is to be used as a permissible value for the rail temperature. A 10 °F difference between ΔT_{Bmax} and ΔT_{Bmin} would generally assure such a barrier.
- If the difference between ΔT_{Bmax} and ΔT_{Bmin} is less than 10 °F (i.e., when buckling energies are small), a safety factor (SF) must be used on ΔT_{Bmin} to provide no buckling potential (since only the prebuckled, stable equilibrium configurations can exist). The case of progressive buckling requires a special treatment as indicated below.
- The determination of ΔT_{Bmax} and ΔT_{Bmin} must be based on a theoretically sound and validated dynamic buckling theory properly accounting for all the key parameters referred to earlier.

In line with these requirements, the safety criterion for the allowable temperature increase is determined as:

$\Delta T_{all} = \Delta T_{Bmin}$;   for $(\Delta T_{Bmax} - \Delta T_{Bmin}) > 10$ °F

$\Delta T_{all} = \Delta T_{Bmax} - SF$;   for $(\Delta T_{Bmax} - \Delta T_{Bmin}) < 10$ °F, and SF is between 0 and 10 °F

The ΔT_{all} for the special case of progressive buckling (Figure 2-3) when $\Delta T_{Bmax} = \Delta T_{Bmin} = \Delta T_{P}$ is:

$\Delta T_{all} = \Delta T_{P} - 10$ °F

The inherent assumption for the progressive buckling case (typical to weak, high degree curve tracks) is that, at the corresponding deflections, ΔT_{P} are not large enough to produce train derailments at low speeds. In the CWR-SAFE program as discussed later, these calculations are performed automatically, and the results are displayed to the user during the safety analysis phase of the program. Figure 2-12 illustrates the safety criterion. On the basis of this work, the UIC has adopted a similar safety criterion.

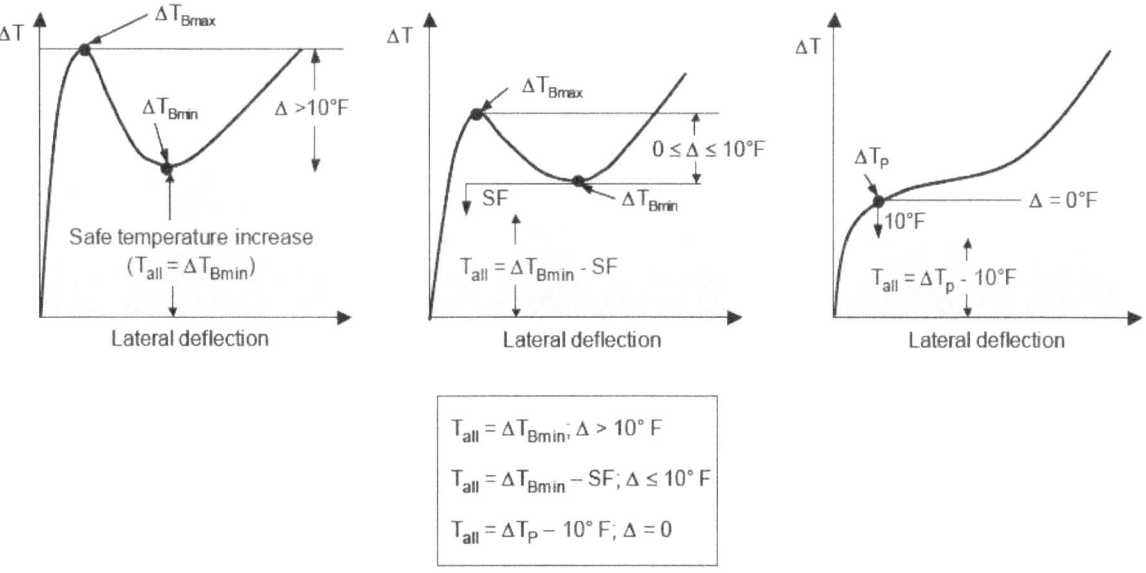

Figure 2-12. Illustration of Safety Criterion

27

A useful safety concept for buckling safety evaluations is the buckling safety margin (BSM), which is a measure of the reserve buckling strength for a given set of track and vehicle parameters. BSM is based on how close the track is to the safety criterion's ΔT_{all} for its specific neutral temperature and maximum rail temperature condition [i.e., the difference between the allowable temperature increase (ΔT_{all}), and the rail temperature (T_R) and the neutral temperature (T_N) difference]

$$BSM = \Delta T_{all} - (T_R - T_N)$$

Four ranges of the BSM identified for practical evaluation of reserve buckling strength exist, and ideally, the BSM should be as large as possible to provide the greatest margin of buckling safety. BSM is automatically calculated in the safety analysis phase of the CWR-BUCKLE program; evaluating the four ranges for the BSM include the following:

- No Margin – In this case, BSM as calculated by the formula above is less than zero. This implies that no margin of safety exists for the given set of parameters, rail temperature, and neutral temperature, and buckling potential is high.
- Minimum Required Range – In this case, BSM is between 0 °F and 20 °F. This implies that only a small margin of safety exists for the track in question.
- Adequate Range – In this case, BSM is between 20 °F and 40 °F, implying that an adequate margin of safety exists for the track in question.
- Desired Range – In this case, BSM is greater than 40 °F. This is considered the desired high range because it provides the largest margin of safety for the given track.

As an example, refer to the buckling response shown in Figure 2-11, the allowable temperature increase, $\Delta T_{all} = 82$ °F. If the maximum rail temperature in that region is 140 °F and if the CWR neutral temperature is 75 °F, BSM is 17 °F, satisfying the minimum required range. This also means that, should the neutral temperature drop down to 58 °F, the track will have a zero BSM and will be highly buckling prone.

2.10.3 Buckling Safety Evaluation Methodology

A methodology for the buckling safety evaluation of CWR track segment consists of a five-step process illustrated in Figure 2-13.

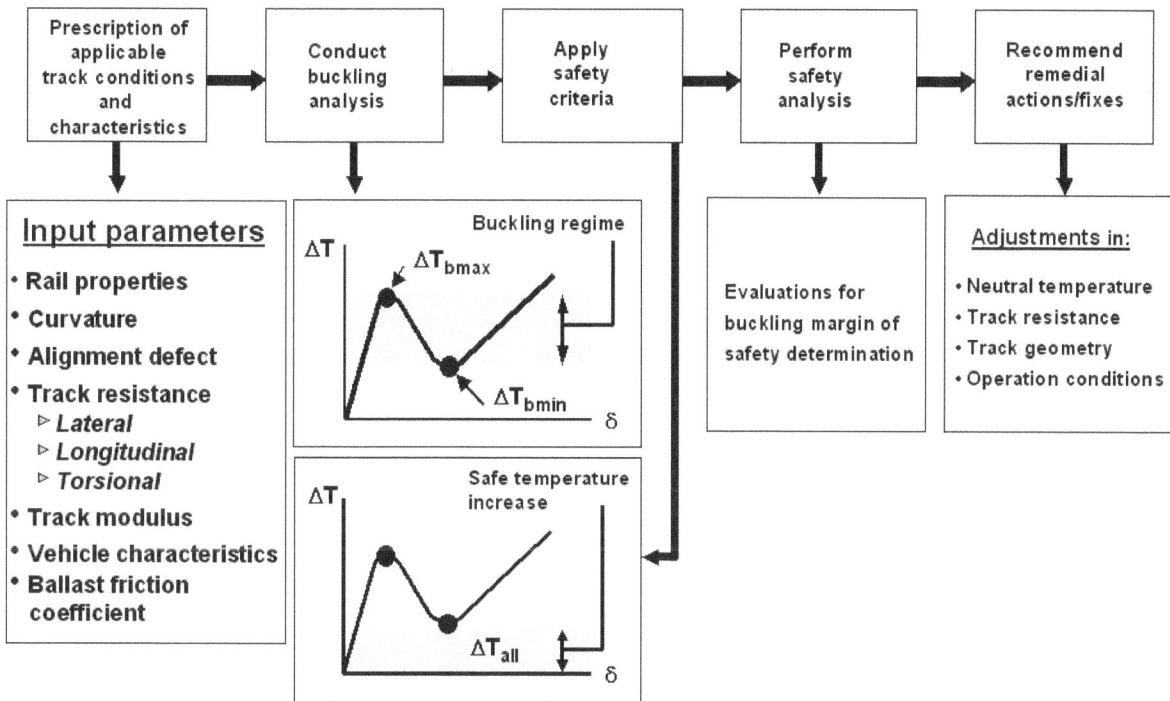

Figure 2-13. CWR Track Buckling Safety Evaluation Methodology

- For the specific track segment under consideration, determine the applicable track input parameters described in Section 2.2.5.
- Conduct buckling analysis to determine the critical temperatures ΔT_{Bmax}, ΔT_{Bmin}, or ΔT_P.
- Apply the safety criterion as described in Section 2.3.2 to determine ΔT_{all}.
- Perform the safety analysis to determine BSM using inputs of rail neutral temperature and maximum rail temperature.
- Determine the range of safety (i.e., whether it is in the no margin, minimum, adequate, or desired range.
- Depending on the specific application or risks involved, decide if the buckling margin of safety is adequate or not.
- If BSM is found to be zero or less, adjustments to the track must be made to reduce the buckling risk.

To increase BSM to a desired level, some of the track parameters/characteristics may need to be changed. The following subsection describes the key parameters which provide the most effective way for increasing BSM. Chapter 4, which deals with parametric studies, will provide additional information for increasing ΔT_{all} and BSM.

2.10.4 Key Controlling Parameters

Track Lateral Resistance

Lateral resistance can be increased by traffic or dynamic consolidation or other compaction means. Ballast section can improve lateral resistance through a wider ballast shoulder and fuller crib level. Alternative tie designs (different tie material, larger tie end cross sections, heavier mass, and better tie bottom and side friction resistance) could be considered for new or rebuilt tracks.

29

Lateral Alignment

A degraded lateral alignment condition can reduce BSM because ΔT_{all} is reduced. An improved lateral alignment will increase the ΔT_{all}. Hence, keeping track to good alignment geometry, through appropriate realignment, and monitoring curves for lateral movement will increase buckling safety.

CWR Neutral Temperature

The railroads strive to maintain a high neutral temperature, typically between 90 °F and 115 °F, depending on regional temperature variations. CWR is known to move longitudinally through fasteners and also laterally in the case of curved tracks resulting from vehicle lateral loads and rail temperature fluctuations. The neutral temperature can fall to 60 °F or lower because of rail and track movement and maintenance actions (curve realignment or rail repair) as found in many tests [15, 19]. Hence, the neutral temperature drop alone can reduce BSM by as much as 30 °F or more. Although at present no convenient way of measuring the rail neutral temperature exists, the changes in neutral temperature can be monitored with basic force measuring strain gages. Timely destressing of CWR to recover the loss in the initial neutral temperature will be helpful in maintaining a high BSM, as well as making effective broken rail repairs, and exercising quality control on the CWR installation and relay processes to ensure a highly desired target neutral temperature.

2.11 Summary

This chapter presented the theory of track buckling. The theory considers the vehicle and thermal loads on tangent or curved CWR track with lateral misalignments and nonlinear lateral resistance characteristics. Appendix A presents the actual mathematical formulations. This chapter also presented the buckling mechanism, assumptions in the development of the theory, and procedures used to determine the buckling strength (in terms of critical forces and temperatures), along with a fundamental safety criterion for buckling prevention and the key parameters influencing it.

3. Track Buckling Model: CWR-SAFE

Chapter 3 presents the fundamental theory of lateral track buckling and the parameters involved. The theory is computerized in a program called CWR-BUCKLE. This program accepts the parameters as they are measured with no simplifying assumptions. The program provides a scientific tool for the evaluation of buckling strength.

The need to specify all the required parameters in CWR-BUCKLE can at times be difficult, especially in direct application by the industry; therefore, a simpler program called CWR-INDY has also been developed. This program requires very simple basic inputs, such as track type, ballast type, crib and shoulder levels, and million gross tons (MGT). From these simplified inputs, the true scientific parameters required for CWR-BUCKLE are estimated through empirical relationships built into the program, and the buckling response is determined using CWR-BUCKLE's computational engine.

CWR-BUCKLE and CWR-INDY represent deterministic methods of predicting the CWR buckling, as opposed to the probabilistic approach in which the buckling is expressed in probabilistic terms, based on statistical variations of key parameter's. Track parameters, such as the lateral resistance, can vary from tie to tie in field conditions. Variations and uncertainty are also associated with other parameters, such as misalignments and neutral temperature.

The probabilistic method overcomes the problem of specifying the parameters precisely and sometimes conservatively, and lends itself to a risk-based analysis and mitigation approach to track buckling. The probabilistic approach is available in the form of a computer program called CWR-RISK. This program takes the statistical parameters as inputs, and the outputs are given in probabilistic terms. The computational engine used for the buckling calculation is again CWR-BUCKLE.

The three modules, CWR-BUCKLE, CWR-INDY, and CWR-RISK, are combined into a single computer program, CWR-SAFE, which can be exercised by different users for both deterministic and probabilistic assessment of CWR track buckling potential.

3.1 Basis of CWR-SAFE Modules

The following sections describe the basis of the three modules in CWR-SAFE.

3.1.1 CWR-BUCKLE

Because CWR-BUCKLE is a deterministic method, it is based on mechanistic buckling theory as explained in Chapter 2 and Appendix A. It was validated by direct test evaluations in field conditions on several tracks on revenue lines and also at the U.S. Department of Transportation's TTC in Pueblo, CO, where static and dynamic buckling tests were carried out during 1983–1984 and 1986–1987 on tangent and curved CWR tracks; test results are available in references [5–8]. To briefly highlight some of validation aspects, the following examples illustrate some key features and findings.

Uplift Wave Influence: Comparison of Buckling Strength under a Hopper Car and a Locomotive

To compare the relative influence of the central bending wave as produced by a loaded 100-ton hopper car and a locomotive, equal levels of misalignment were set under each of the vehicles.

Vertical and lateral displacements under the stationary cars were measured as the rails were heated. Figure 3-1 shows a comparison of lateral displacements under each vehicle as a function of temperature. The misalignment growth under the hopper car was found to be more severe, indicating the influence of the longer uplift wave under the hopper car. The tests also showed that the uplift wave is a contributing factor in the misalignment growth mechanism, hence an important component of the dynamic buckling analysis. This test helped better understand the importance of the vehicle characteristics, such as the truck center spacing and vertical axle loads, their influence on track uplift, and the need to incorporate this influence in the buckling model.

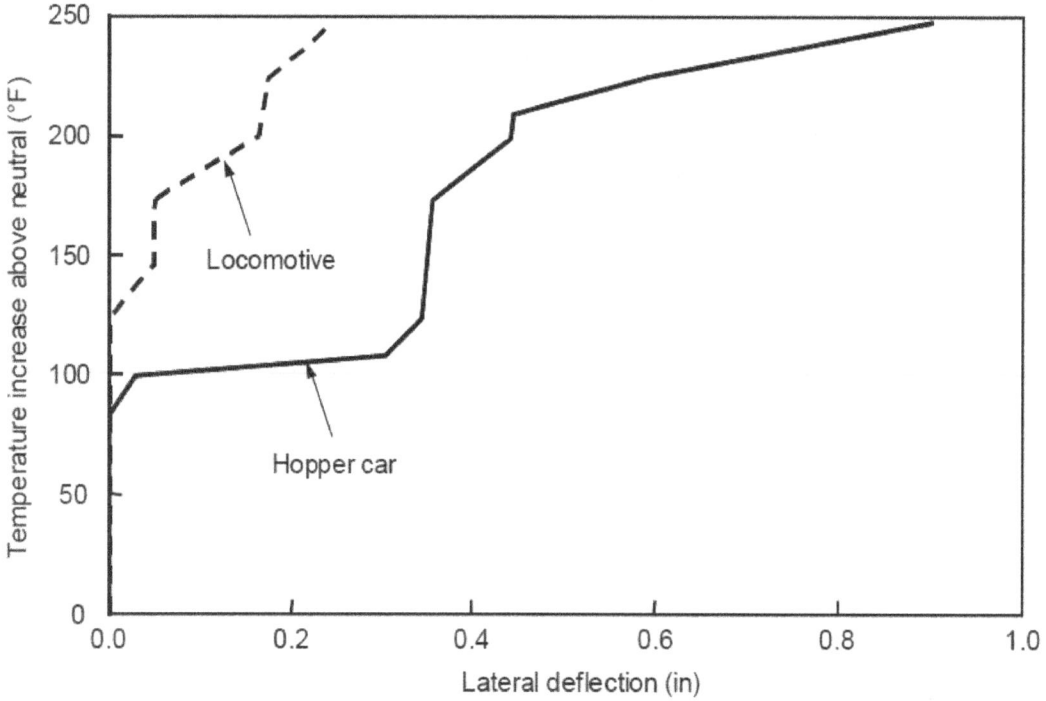

Figure 3-1. Response of Track under Vehicles

In another test, the measured response of the track with a large misalignment under a stationary hopper car favorably compared with the theoretical prediction, as shown in Figure 3-2, giving validation of model's predictions when the buckling response is progressive.

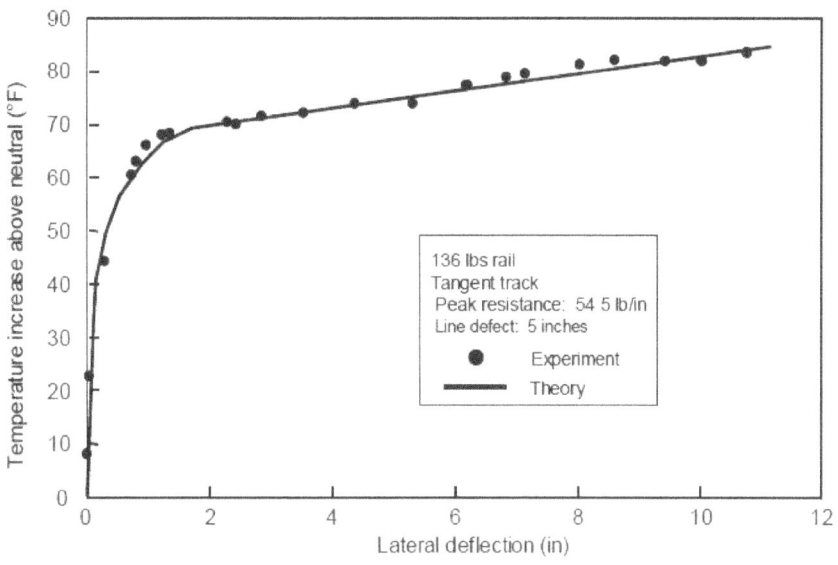

Figure 3-2. Progressive Buckling Test (Tangent)

Dynamic Buckling Response Validation

A weak 5-degree curved track was tested dynamically using a locomotive and hopper car consist making several passes at slow speeds. After reaching 40 °F with five incremental train passes, no growth of initial misalignment resulted. Train passes made above 40 °F increased the initial misalignment to 1.1 in; at 62 °F, the curve buckled to a deflection of approximately 8 in after the 10th pass, as shown in Figure 3-3. The measured dynamic buckling response was in good agreement with the theoretical predictions of CWR-BUCKLE.

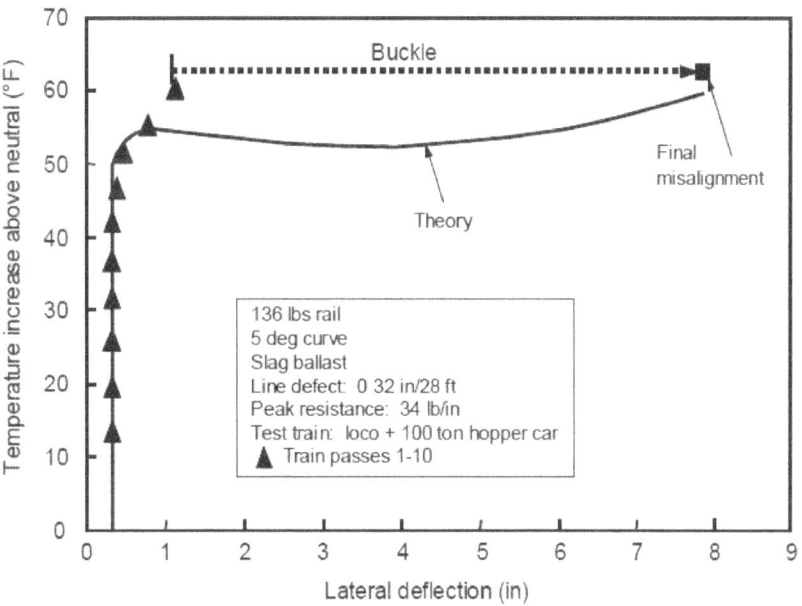

Figure 3-3. Dynamic Buckling of Curved Track

Effect of Car Length/Uplift Wave on Misalignment Growth

To further study the uplift wave influence on the lateral response, the growth of imperfections under the passage of different cars was monitored using strip chart recorders. Figure 3-4 is a snapshot from one of the charts. The intent was to evaluate the hopper car's larger uplift wave influence on helping the buckling mechanism by producing larger misalignments versus a locomotive. An incremental growth under the hopper car versus a nonincreased deflection under the locomotive further proved the need for uplift based vehicle buckling theory used in the CWR-BUCKLE program.

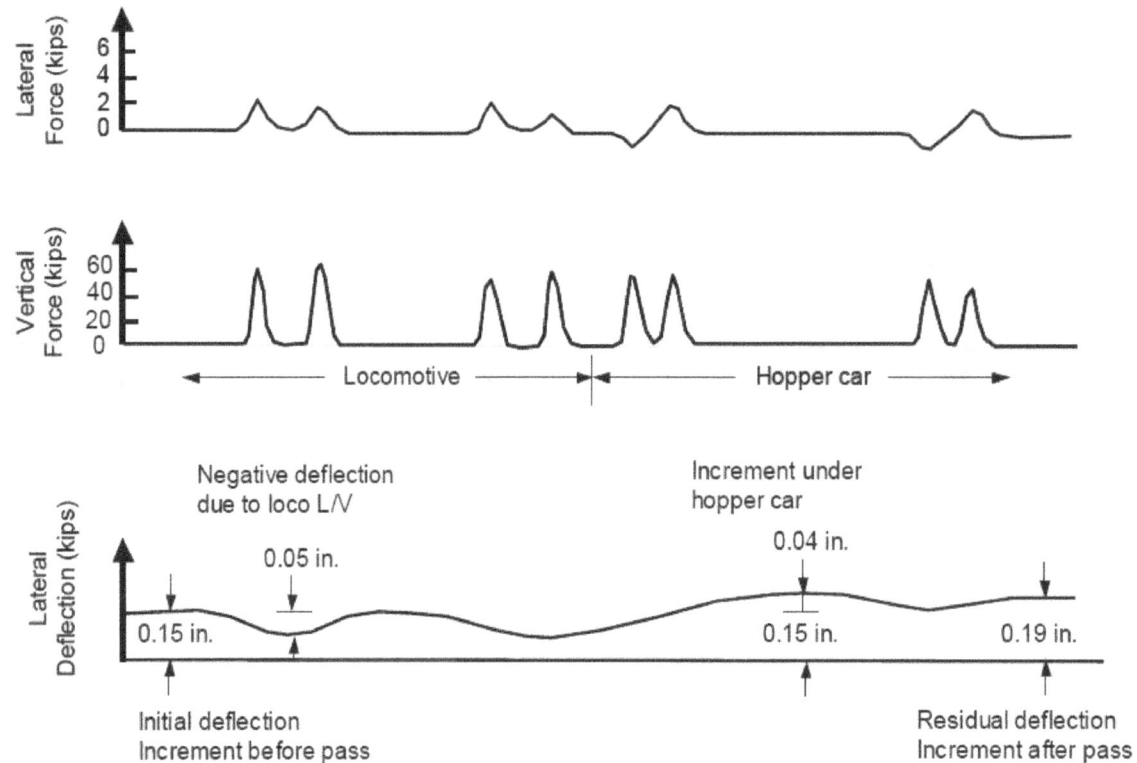

Figure 3-4. Strip Chart Record for 5-Degree Curve Test

The foregoing test examples illustrate some of the test validation aspects and, coupled with other tests and studies, provide adequate validation of the uplift based buckling theory of CWR-BUCKLE.

CWR-BUCKLE Features

CWR-BUCKLE is a scientific computer program, utilizing certain input parameters of track and vehicle, to predict the following:

- The buckling strength, expressible in terms of the two critical temperatures, T_{Bmax} and T_{Bmin}, or the T_P as defined previously in Chapter 2. It can predict the entire pre- and postbuckling response curve of the temperature versus lateral deflection.
- It can compute the buckling energy of the track at a given rail temperature.
- It predicts the rail forces before and after buckling.

34

- It predicts the allowable rail temperature increase for buckling safety and the track's BMS. The BMS is indicative of the reserve buckling strength in terms of additional acceptable temperature increase before buckling. If the margin of safety is not adequate (Chapter 2), the program provides recommendations to the user on how to improve BMS.

The following example illustrates a typical CWR-BUCKLE run with the input parameters shown in Table 3-1.

Table 3-1. Input Parameters for CWR-BUCKLE Illustrative Example

CWR-BUCKLE Illustrative Example Inputs

Rail size (lb/yd):	AREA 136
Tie type:	Concrete
Tie weight (lb):	750
Tie spacing (in):	24
Track curvature (deg):	Tangent
Ballast type:	Granite
Tie ballast friction coefficient:	0.86
Torsional resistance (in-kips/rad/in):	25
Longitudinal stiffness (psi):	200
Foundation modulus (psi):	10000
Peak lateral resistance (lb/in):	150
Misalignment amplitude (in):	1.5
Misalignment half-wavelength (in):	180
Rail neutral temperature (°F):	50
Maximum rail temperature (°F):	135
Vehicle type:	Hopper

With the above inputs, CWR-BUCKLE provides a graph displaying buckling temperature and deflection as shown in Figure 3-5. The figure shows the two salient temperatures, namely the ΔT_{Bmax} = 172 °F and ΔT_{Bmin} = 90 °F. In addition to the graph, the output includes a buckling results summary and a safety analysis, as shown in Table 3-2 and Table 3-3.

35

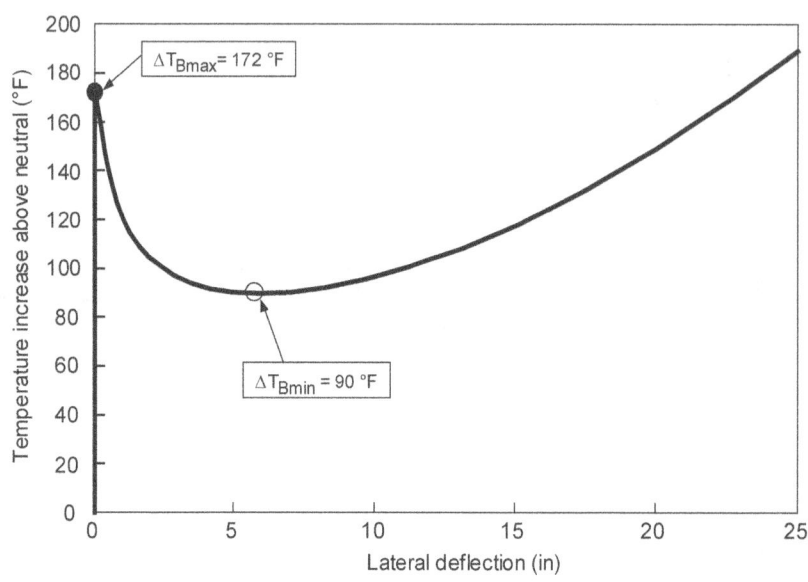

Figure 3-5. Buckling Deflection versus Temperature Increase above Neutral Temperature

Table 3-2 shows that if the track were to buckle out at the upper critical temperature of 171.66 °F, the corresponding amplitude would be 22.97 in with the half wavelength of 235.26 in. The force in the rail to cause the buckle is 439.96 kips, and at the ΔT_{Bmax} of 171.66 °F, the required energy to buckle the track is zero (as opposed to the energy required to buckle the track at the lower critical temperature ΔT_{Bmin} = 89.80 °F, which is 8.47 in-kips).

Table 3-2. CWR-BUCKLE Output Buckling Results Summary

```
        The upper and lower critical buckling temperature increases,
    deflections, wavelengths, forces and energies for this case
    are as follows:

At the LOWER CRITICAL TEMPERATURE INCREASE,

Temperature increase above neutral   =  Tb,min  =     89.80 F
Max buckling deflection              =  Wmax    =      5.95 in
Buckled half-wavelength              =  L       =    190.00 in
Force in rail at time of buckle      =  P       =    230.17 kips/rail
Energy required to buckle            =  E       =      8.47 in-kips

At the UPPER CRITICAL TEMPERATURE INCREASE,

Temperature increase above neutral   =  Tb,max  =    171.66 F
Max buckling deflection              =  Wmax    =     22.97 in
Buckled half-wavelength              =  L       =    235.26 in
Force in rail at time of buckle      =  P       =    439.96 kips/rail
Energy required to buckle            =  E       =       .00 in-kips
```

Table 3-3 shows the safety analysis that gives the safe allowable temperature increase, T_{all} = ΔT_{Bmin} = 89.8 °F, and, if the neutral temperature for this case is 50 °F, then the track should not buckle until T_r = 89.8 + 50 = 139.8 °F. Because the maximum rail temperature in this example was set at 135 °F, the BSM is approximately 5 °F.

36

Table 3-3. CWR-BUCKLE Output – Buckling Safety Analysis

```
        For this case, the difference between the upper
  and lower critical buckling temperature increases is
  greater than 10 deg F.  The maximum allowable rail
  temperature for this case is thus given by:


      Tall    =    allowable rail temperature increase above
                   neutral

              =    Tb,min
              =    89.80 deg F

      The results of the buckling safety analysis are
  as follows:

      Tall    =    allowable rail temperature increase above
                   neutral ( 89.80 deg F)
      Tn      =    rail neutral temperature ( 50.00 deg F)
      Tr      =    maximum rail temperature (135.00 deg F)

                   Buckling will occur if:
                   (Tr  -  Tn)  >  Tall

                      For this case,
              (Tr  -  Tn)  =  85.00  <  Tall
  Based on the above, BUCKLING SHOULD NOT OCCUR FOR THIS CASE.

      The Buckling Safety Margin for this case is:   4.80 deg F.
  This Safety Margin is within the MINIMUM REQUIRED range for
  buckling safety.
```

Summary Conclusions

- CWR-BUCKLE is a versatile analysis tool used to perform detailed buckling evaluation for given track and vehicle inputs. Full-scale experiments in the field validated the program.
- In addition to the quantitative analysis on the buckling strength, the program outputs the margin of safety and recommends measures to increase this margin of safety if necessary.

3.2   CWR-INDY

As indicated above, CWR-BUCKLE requires many complex parameters for buckling analysis. As an alternative, the Industry Version, called CWR-INDY, was developed, which requires simpler inputs, such as ballast type and condition, tie type, and fastener type. Similar to CWR-BUCKLE, the model requires neutral and maximum rail temperatures as inputs, and it assumes a vertical modulus of 6,000 psi for wood tie track and 10,000 psi for concrete tie track. With these simpler inputs, the program internally determines the actual numeric values for the required parameters and will use the same core analysis as in CWR-BUCKLE.

This section describes the basis, assumptions, and relationships used in this CWR-INDY version to translate the simpler input characteristics into the actual required technical parameters. The program has been developed to require the input parameters known to the railroad user, which include the following:

- Rail size (AREA 100,115,119,132,133,136, and 140; UIC 54 and 60)
- Tie type (wood or concrete)
- Ballast material (granite or slag)
- Fastener type (cut spikes, or elastic fasteners)
- Anchor/fastener pattern (every tie anchored (ETA) or every other tie anchored (EOTA))

37

- Ballast crib level (full, ¾, or ½)
- Ballast shoulder width (up to 18 in)
- Tamping (yes or no)
- Consolidation level (MGT)
- Dynamic track stabilization (in terms of MGT equivalent)
- FRA track class (for lateral misalignments determination)

On the basis of the foregoing information, the following engineering input parameters are internally determined by the program:

- Lateral resistance
  - Peak value, F_P
  - Limiting value, F_L
  - Displacement w_P at F_P
  - Displacement w_L at F_L
- Torsional resistance
- Longitudinal resistance
- Misalignment amplitude and wavelength

The output of the program is the ΔT_{all}, the margin of safety for expected rail maximum temperature for a given neutral temperature, and the numerical value for the internally computed peak lateral resistance, F_P.

Following is a description of the program's internal algorithms, assumptions, tables and equations to determine the engineering parameters from the design parameter type inputs.

1. Lateral Resistance

The Single Tie Push Test (STPT) (see Appendix B) typically exhibits a characteristic such as shown in Figure 2-8. The technical parameters are peak resistance (F_P), limit resistance (F_L), deflection at peak resistance (w_P), and deflection at limit resistance (w_L).

Because this information is not usually available to railroad personnel, the industry version of the program estimates the lateral resistance from the track characteristics. The required characteristics include the following:

- Tie type
- Ballast type
- Track consolidation
- Track maintenance
- Shoulder width
- Crib level

Basis of Empirical Correlation

The peak lateral resistance of the tie under full ballast conditions can consist of the two components listed below. Empirical formulas based on test data have been developed for these components.

- The resistance component in the tamped condition

38

- The increment in the resistance resulting from traffic and the Dynamic Track Stabilizer (DTS)

The resistance reduction due to a substandard ballast section (not having full crib or full shoulder) is obtained by using a reduction factor (see Equation 3-4) on the fully ballasted condition resistance; this factor is less than 1 and depends on the reduced shoulder and crib levels measured in tests [10,11].

Resistance under Full Ballast

The peak resistance for full ballast (full crib with shoulder between 12 and 18 in or more) can be expressed as:

$$F_P = F_{P0} + \Delta F_P \hspace{4cm} \text{(Equation 3-1)}$$

where F_{P0} is the tamped resistance, and ΔF_P is an increment attributable to traffic consolidation and DTS.

Tamped Resistance F_{P0}

The resistance depends on the type of tie, ballast, and type of tamping. For tamped track, Table 3-4 provides typical F_{P0} values. A range exists for these values, but the values shown here are on the conservative side, according to the test data from many sites [8, 9]. If lift accompanies tamping, subtract 300 from the F_{P0} values on Table 3-4.

Table 3-4. Resistances Tamped Conditions

Ballast Type	Tie Type	F_{P0} (lb)
Granite	Wood	1,900
Slag	Wood	1,650
Granite	Concrete	2,300

Increment because of Consolidation, ΔF_P

From tests [10], Figure 3-6 shows the increment caused by traffic consolidation, expressed as MGT, for wood ties.

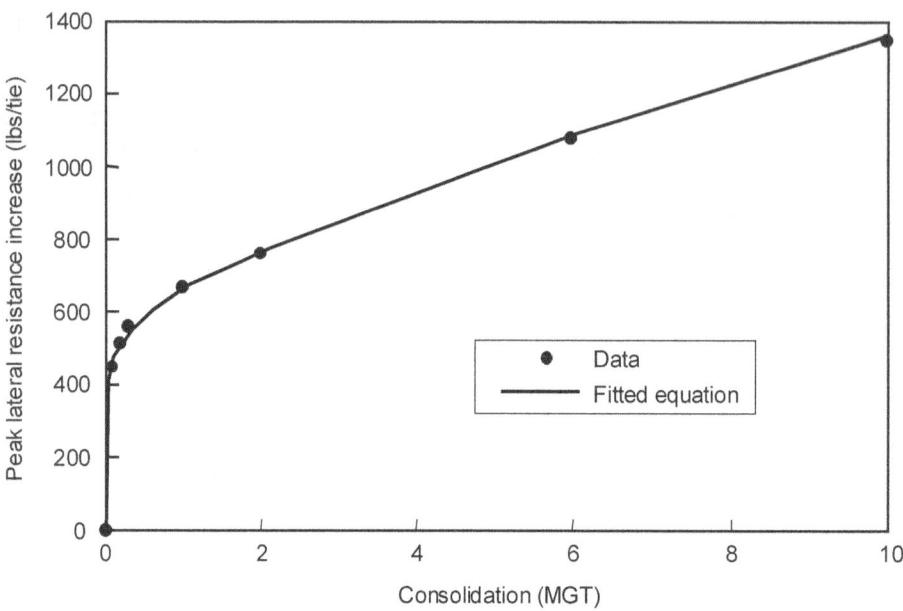

Figure 3-6. Consolidation Influence - Wood Ties for Equation 3-2

The equation best fitting the curve for the wood tie track is the following:

$$\Delta F_p = 60 \bullet MGT + 944 \bullet \left(1 - e^{-MGT^{0.2}} \right) \text{ lbs}; \quad MGT<10 \qquad \text{(Equation 3-2)}$$

Tests have shown that the difference in the resistance increment for slag and granite ballast is not very significant [10]. Hence, Equation 3-2 is valid for slag and granite cases of the wood tie track.

In a similar manner, the increment in concrete tie resistance attributable to consolidation is provided by Equation 3-3 as a good fit to the test data on concrete ties shown in Figure 3-7:

$$\Delta F_p = -4.7 \bullet MGT + 1360 \bullet \left(1 - e^{-MGT^{0.4}} \right) \text{ lbs}; \quad MGT<10 \qquad \text{(Equation 3-3)}$$

(For MGT > 10, the model assumes asymptotic values just exceeding the MGT = 10 values.)

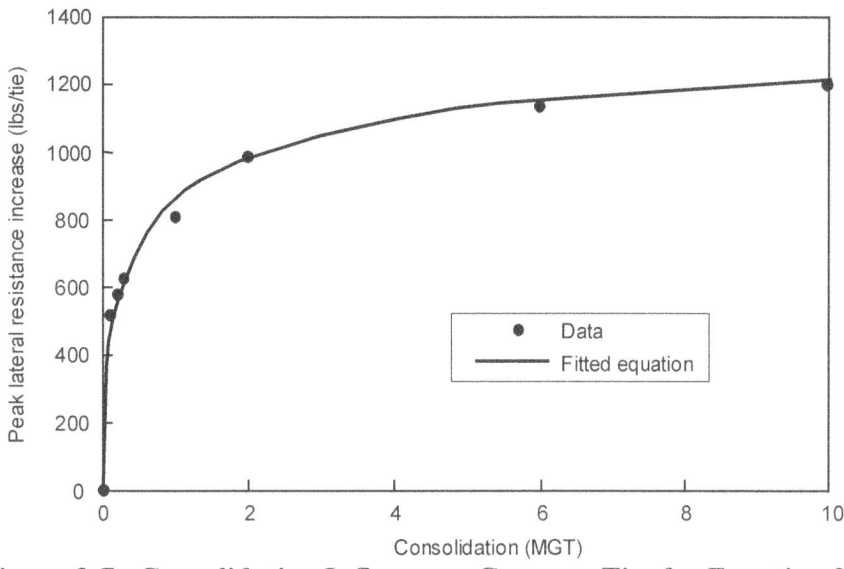

Figure 3-7. Consolidation Influence – Concrete Ties for Equation 3-3

In the case of dynamically stabilized track, the MGT term in the above equation is replaced by MGT + equivalent MGT in dynamic track stabilization (DTE), where DTE is an equivalent MGT for the given type of dynamic track stabilization operation. This equivalent MGT depends on the setting of the track stabilizer (frequency and head pressure), its speed, and the number of passes made. The DTE equivalent suggested for CWR-INDY use is 0.1 MGT based on European tests and UIC Code 720 [20]. This must be confirmed by tests on U.S. tracks for wood and concrete ties on tangent and curved tracks.

Reduction due to Reduced Ballast Levels

Optimum ballast section is one with full cribs and shoulder widths in excess of 12 in. If ballast component levels are reduced, the resistance will decrease, and a multiplication factor is used to adjust for the reduced resistance in CWR-INDY. Research has shown that there is a limiting shoulder width beyond which the resistance is not affected. This is between 18 and 24 in according to reference [21], with the lower value of 18 in being assumed in CWR-INDY. If the ballast shoulder is less than 18 in and the crib level is not full, the lateral resistance will be reduced proportionately using the formula:

$$\text{Reduction Factor} = \left(\beta_1 + \beta_2 C_d + \beta_3 \frac{S_w}{18} \right)$$
(Equation 3-4)

In the foregoing equation, C_d represents the ratio of existing crib depth to the full depth. S_w represents the shoulder width 18 in (maximum) or less. Table 3-5 shows values of β_1, β_2, and β_3. For full crib and 18 in or more shoulder, the factor is one, as seen in Equation 3-4.

Table 3-5. β Values for Equation 3-4

β_i	Wood Tie	Concrete Tie
1	0.30	0.40
2	0.50	0.45
3	0.20	0.15

41

Lateral Resistance Equation

The general equation for F_P becomes:

$$F_P = \left[F_{P0} + \alpha(MGT + DTE) + \gamma(1 - e^{-(MGT+DTE)^\lambda}) \right] \left(\beta_1 + \beta_2 C_d + \beta_3 \frac{S_w}{18} \right) \qquad \text{(Equation 3-5)}$$

where,

F_P is in lb
F_{P0} is from Table 3-4
α is 60 for wood tie track; -4.7 for concrete tie track
γ is 944 for wood tie track; 1,360 for concrete tie track
λ is 0.2 for wood tie track; 0.4 for concrete tie track
MGT is tonnage expressed in millions of gross tons (limited to <100)
DTE is an MGT equivalent for DTS
β is from Table 3-5

Limiting Resistance

The limiting resistances are evaluated using the following empirical formulas based on test data in accordance to reference [9]:

- Wood tie track

  Granite ballast

$$F_L = \begin{cases} 0.3F_P + 500 \text{ lbs} & \text{if } F_P \geq 715 \text{ lbs} \\ F_P & \text{if } F_P < 715 \text{ lbs} \end{cases} \qquad \text{(Equation 3-6)}$$

  Slag ballast

$$F_L = \begin{cases} 0.06F_P + 600 \text{ lbs} & \text{if } F_P \geq 638 \text{ lbs} \\ F_P & \text{if } F_P < 638 \text{ lbs} \end{cases} \qquad \text{(Equation 3-7)}$$

- Concrete tie track

$$F_L = \begin{cases} 0.38F_P + 950 \text{ lbs} & \text{if } F_P \geq 1532 \text{ lbs} \\ F_P & \text{if } F_P < 1532 \text{ lbs} \end{cases} \qquad \text{(Equation 3-8)}$$

Deflections

The deflections corresponding to F_P and F_L (see Figure 2-8) are also set in the CWR-INDY with the following values based on data from references [10, 11]:

- Wood tie track

  $w_P = 0.3$ inches
  $w_L = 0.025 \bullet F_P + 2.6$ inches   (granite ballast)
  $w_L = 0.009 \bullet F_P + 3.5$ inches   (slag ballast)

- Concrete tie track

  $w_P = 0.25$ inches
  $w_L = 2.25$ inches

2. *Torsional Resistance*

The torsional resistance versus the angle of rotation behavior between the ties and the rail is linearized based on the test data [10]. Table 3-6 shows the conservative torsional stiffness values for some frequently used fasteners.

Table 3-6. Torsional Stiffness, τ_0 (for two rails)

Tie Spacing (inch)	Fastener Type	τ_0 (in-kips/rad/in)
20	Cut Spikes on Wood	80
20	Pandrol on Wood	370
24	Pandrol on Concrete	10
24	McKay on Concrete	25

3. *Longitudinal Resistance*

Track longitudinal resistance is the resistance offered by anchors, fasteners, ties, and ballast to the rails as they tend to move in the longitudinal direction as a result of thermal force gradients, train braking, and traction forces and during buckling. This resistance is assumed to be proportional to the rail longitudinal displacement, hence defined through a stiffness parameter.

The longitudinal stiffness depends on the fastener type and ballast condition. Table 3-7 presents the values assumed in the program for wood tie anchors, Pandrol fasteners on wood or concrete ties, and McKay fasteners on concrete ties. The values shown are for tight anchors. For degraded or loose anchors, the assumed values are half of those shown in Table 3-7. The wood tie data presented are from the tests conducted at TTC, as reported in reference [9]. The concrete tie fastener resistance data comes from unpublished test data from TTC. These values are considered to be conservative. A loose anchor resistance program assumes half the value for anchors.

Table 3-7. Longitudinal Stiffness (for two rails)

Anchoring or Fastener	k_f (lb/in/in)	
	Tamped	Consolidated
ETA on Wood (tight anchors)	200	400
EOTA on Wood (tight anchors)	100	200
Pandrol on Wood	200	400
Pandrol on Concrete	200	400
McKay on Concrete	200	400

4. *Misalignment Parameters*

Misalignment amplitudes are internally set in accordance with the FRA class of track specified by the user, in accordance with FRA's *Track Safety Standards Compliance Manual* limits. The program internally calculates the corresponding wavelength through a track lateral deformation algorithm based on pushing the track laterally to a deflection level equal to the FRA class misalignment amplitude specified and evaluating its corresponding wavelength.

43

- The inputs to CWR-INDY program for CWR buckling analysis are simpler than those for CWR-BUCKLE. The parameters are internally estimated from simple inputs such as the ballast crib and shoulder levels, types of track, and MGT.
- With the simple inputs converted into their equivalent scientific parameters, the CWR-INDY program uses CWR-BUCKLE as the computational engine to perform buckling analysis.
- The output of the program also includes the calculated peak lateral resistance value since this was not directly required as an input. In this sense, CWR-INDY can also function as a lateral resistance predictor. The output gives similar information as CWR-BUCKLE discussed in the previous subsection; however, provided as part of the output is graphical display of temperature versus lateral deflection.

Numerical Example

To illustrate how CWR-INDY works, this section presents a numerical example with input parameters similar to those used in the CWR-BUCKLE example, shown in Table 3-8.

Table 3-8. Input Parameters for INDY Illustrated Example

CWR-INDY Illustrative Example Inputs	
Rail size (lb/yd):	AREA 136
Tie type:	Concrete
Tie weight (lb):	750
Tie spacing (in):	24
Track curvature (deg):	Tangent
Ballast type:	Granite
Fastener type:	Pandrol
Crib level:	Full
Tamped:	NO
Stabilized:	NO
Track consolidation (MGT):	1
Shoulder width (in):	12
Track class:	4
Rail neutral temperature (°F):	50
Maximum rail temperature (°F):	135

First, the program calculates a value for T_{all}, which in this example is 92 °F. The neutral temperature is given as 50 °F, meaning that the risk of buckling occurs at 142 °F. As in CWR-BUCKLE, the input maximum temperature is 135 °F. CWR-INDY calculates the BSM as 7 °F, which is the difference between the maximum rail temperature and T_{all}. Table 3-9 displays the actual output shown by CWR-INDY. CWR-INDY also displays the calculated peak lateral resistance of 3,133 lb, which was obtained for the input parameters of concrete tie, granite ballast, 12-inch shoulder, full cribs, and 1 MGT of consolidation.

Table 3-9. CWR-Indy Output – Buckling Safety Analysis

```
Based on the input parameters, the estimated peak lateral
resistance per tie, Fp, is 3133 lbs.  You are advised to
check this value through Single Tie Push Tests (STPT) on a
few ties in the subject track.


The results of the buckling safety analysis for the
estimated lateral resistance are as follows:


Tall  =  allowable rail temperature increase above
         neutral ( 92 deg F)
Tn    =  rail neutral temperature ( 50 deg F)
Tr    =  maximum rail temperature (135 deg F)


         Buckling is likely to occur if:
         (Tr - Tn) is greater than Tall


              For this case,
(Tr - Tn) = 85 deg F, which is less than Tall

Based on the above, BUCKLING SHOULD NOT OCCUR FOR THIS CASE.

The Buckling Safety Margin for this case is:  7 deg F.
This Safety Margin is within the MINIMUM REQUIRED range for
buckling safety.
```

Summary Conclusions

- The inputs for the CWR-INDY version are simpler than those for CWR-BUCKLE. The inputs are qualitative (tie type, ballast type, and track maintenance) and quantitative (consolidation level, ballast crib level, and shoulder width).
- Empirical formulas based on the available test data and engineering judgment have been developed to determine the scientific parameters required for CWR-BUCKLE from the CWR-INDY program inputs.
- The empirical formulas account for the effects of tamping, dynamic track stabilization, and consolidation by traffic on the track lateral resistance. Simple conservative values are proposed for the torsional and longitudinal resistance of commonly used fasteners based on test data. A detailed user's guide for CWR-INDY as a part of CWR-SAFE is available in the self-help menu of the software.

3.3   CWR-RISK

The buckling safety analyses performed in CWR-BUCKLE and CWR-INDY can be considered as deterministic analyses because all the input parameters have definite values so that the track either buckles or does not. A track safety/maintenance strategy based on such deterministic analyses can be conservative and therefore expensive because it must be based on the worst-case scenario for all parameters. It is expedient, therefore, to use a probabilistic approach, which can account for the inevitable statistical variations in input parameters. Such methodology will also provide improved flexibility in determining maintenance options and performing safety evaluations. As an example, with a risk-based approach, one can vary and select a range of values for the ballast condition and change the various CWR neutral temperature options to achieve similar levels of buckling safety. Such a choice permits more optimum allocation of maintenance resources.

The use of probabilistic approaches is the current trend in other technology areas requiring structural failure evaluations and safety assessments. The nuclear, aircraft, and naval industries have long benefited from such methods, which are extendable to railroad applications,

45

particularly to developing probabilistic estimates of buckling failures. The risk methodology for this purpose requires not only the failure probability but also the severity or the consequence of the failure. For example, if buckling is predicted, does it cause a derailment and with what damage level?

As applied to track buckling-induced derailments, a slow speed coal freight train operating in a high-degree curve may not result in the same level of damage as a high-speed corridor passenger train on a tangent track. Tangent track tends to buckle explosively with a large deflection, whereas the curved track may buckle progressively with comparatively smaller buckle amplitudes. Even if the parameters of the tangent and the curved track are such that they give equal probability of buckling, the severity of the passenger vehicle on the buckled tangent can be catastrophic compared with that of the freight car on the buckled curve. Therefore, the overall risk of buckling must be measured by the probability of the event occurring, weighted by the severity of the consequence or damage caused by that event.

In the present version of CWR-RISK, attention is focused on buckling probability. The severity aspects of the risk methodology are not treated in detail, but an assumption is made that the severity is proportional to the square of the train speed, which provides a rationale for railroads to use a slow-order policy when the potential exists for rails to buckle at high temperatures.

Buckling Probability Basics

The buckling load will be expressed in terms of the rail temperature increase over the neutral temperature, and the strength is expressed in terms of the allowable temperature increase, ΔT_{all}.

Thus,

$$(\Delta T)_{Load} = T_r - T_n \qquad \text{(Equation 3-9)}$$

$$(\Delta T)_{Strength} = \Delta T_{all} \qquad \text{(Equation 3-10)}$$

The fundamental parameters in the evaluation of failure probability of a structure are load and strength, both of which vary probabilistically in service life of the structure. The intersecting or interference zone in this type of graph represents the temperature regime in which the load equals or exceeds the strength. The probability of this load exceeding the strength is the failure probability of the structure (see hatched area in Figure 3-8). It can be evaluated on the basis of the convolution integral, as discussed in Appendix C.

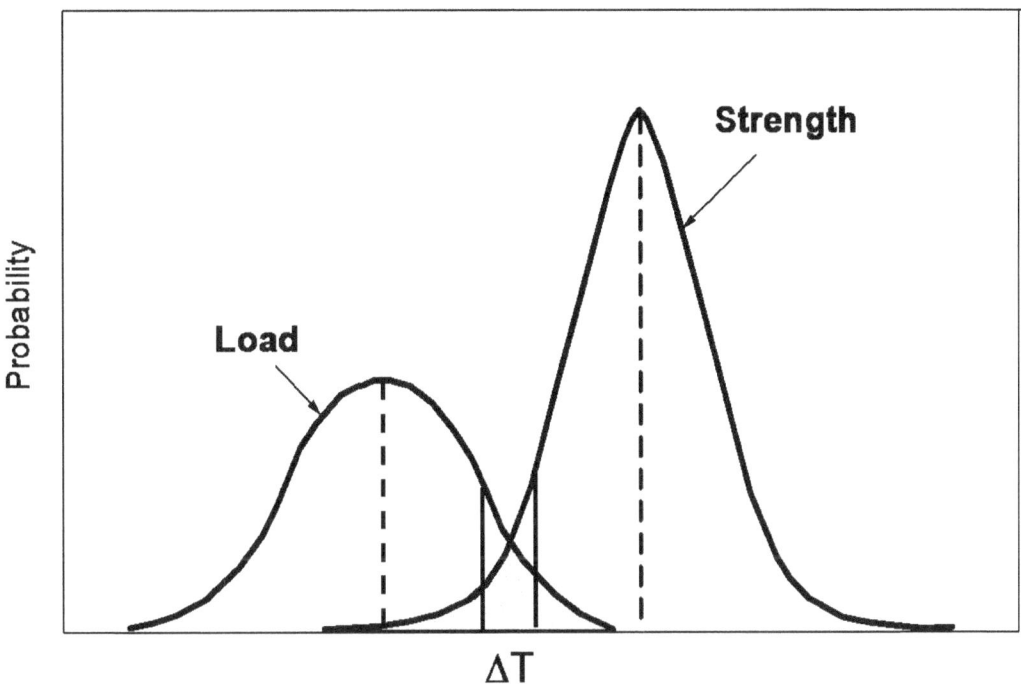

Figure 3-8. Schematic of Interference Principle

The allowable temperature increase ΔT_{all} is determined for the following statistically varying primary track parameters:

- Lateral resistance
- Lateral misalignment

Other parameters, such as the torsional and longitudinal resistance and vehicle loads, are also important in the buckling strength assessment. Their influence is generally small when compared with that of the two primary parameters (see Chapter 4); hence, their variations in the field conditions are not accounted for in CWR-RISK. Figure 3-9 shows these and other secondary parameters, which will be treated as having prescribed deterministic values.

The load probability is obtained from the neutral temperature probability and the rail temperature under consideration. The load is expressed as $(T_r - T_n)$, and only the positive part representing the compressive load is important since tensile load cannot cause buckling. Figure 3-9 shows the overall approach. The strength probability is calculated for the statistical variations of the two primary parameters, namely the lateral resistance and misalignments. In CWR-RISK, CWR-BUCKLE is also used as the computational engine, as discussed in Chapter 5.

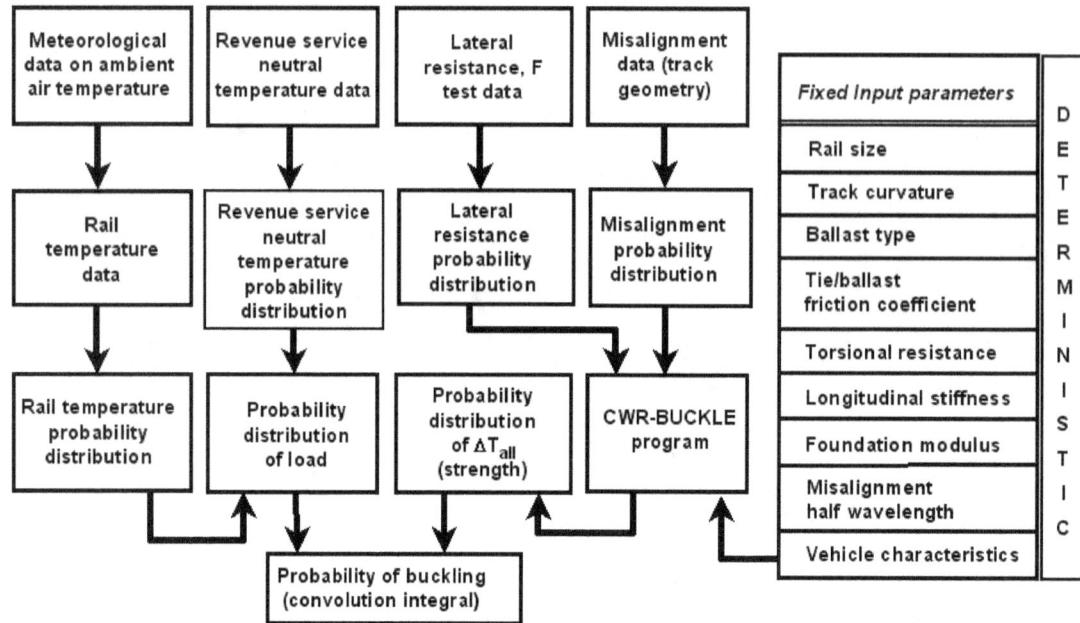

Figure 3-9. Buckling Risk Evaluation Methodology

The rail temperature is an important parameter in the load quantification. From a practical point of view, this represents a critical factor for the railroads on slot timing decisions for CWR track maintenance, and slow orders. With the probabilistic descriptions of lateral resistance, misalignment, and rail neutral temperature, CWR-RISK calculates the probability of buckling at a given rail temperature. The following will first describe the statistical inputs required in CWR-RISK.

Statistical Input Parameters for CWR-RISK

In the following paragraphs, the three important input parameters, (i) rail neutral temperature, (ii) lateral resistance, and (iii) lateral misalignment amplitude, will be described in statistical terms. There is some uncertainty associated with the parameters that vary along the track. It is assumed that a sufficient number of measurements are made over an appropriate length of track segment to characterize the parameters. The number of measurements and segment length are to be established through properly designed trial studies and experiments. Appendix D gives recommendations as to the minimum number of measurements required.

Rail Neutral Temperature Frequency

CWR installation temperature is typically set within a desired range established in the railroad's procedures and is typically between 90 °F and 110 °F, depending on the geographic location. This installation temperature is the rail initial neutral temperature. However, as research has shown [12, 14], the neutral temperature does not stay at the rail installation temperature and can decrease to lower values, as low as 50 °F in some cases. These reductions generally result from rail and track movements (creep, curve breathing, track settlement) and track maintenance activities. The frequency distribution in this range is not expected to be a normal distribution, although this has not been evaluated. For the purpose of numerical illustration, Figure 3-10 shows the assumed distribution based on limited U.S. testing and field experience.

48

The neutral temperature considered here is the average of the two rails. The distribution in Figure 3-10 is spatial. Although the neutral temperature can vary with time after CWR installation or rewelding and trafficking, it is assumed that a steady-state value exists at a given location. Figure 3-10 can be constructed on the basis of one-time (preferably spring) testing at a number of locations spread over the territory. Methods are available to determine the neutral temperature, such as through pre- and postcutting strain gage measurements or rail uplift type methods. It is, therefore, possible to develop a database on the neutral temperatures at different locations to determine the frequency as shown in Figure 3-10. The CWR-SAFE User's Guide provides additional information on the binning and construction of frequency distributions such as in Figure 3-10, which is based on measurements.

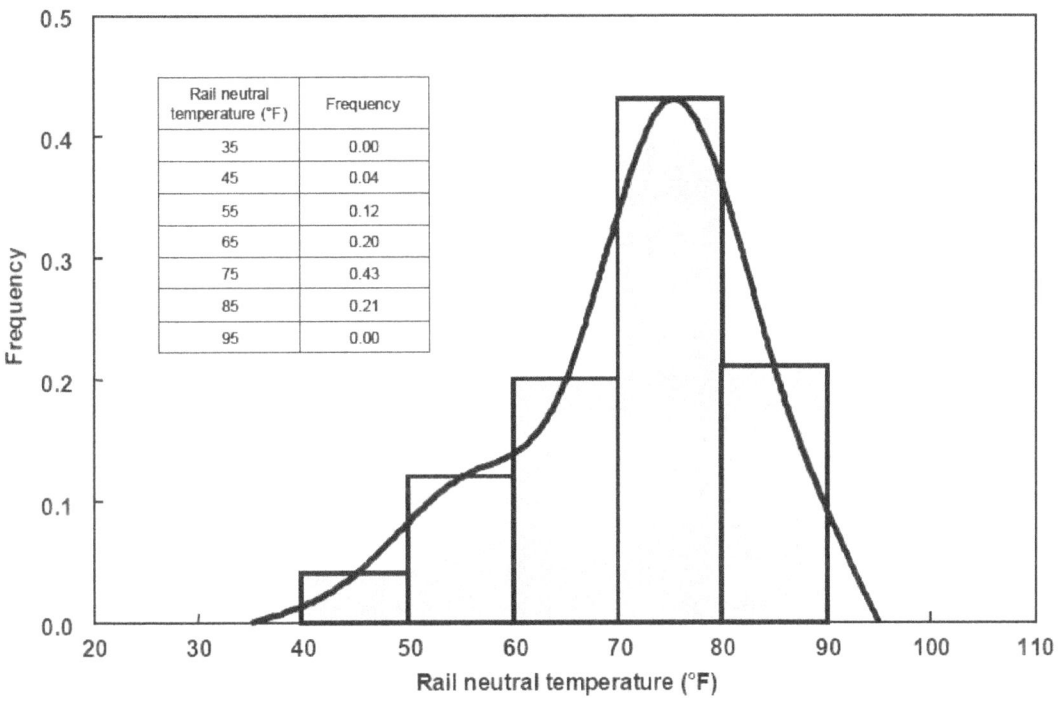

Rail neutral temperature (°F)	Frequency
35	0.00
45	0.04
55	0.12
65	0.20
75	0.43
85	0.21
95	0.00

Figure 3-10. Assumed Rail Neutral Temperature Distribution

Lateral Resistance Frequency

Lateral resistance varies along the track for a given track type and condition. Scatter in lateral resistance values along the track is inherent in the nature of the railroad environment because of varying cribs and shoulders, recent maintenance, tie lateral and longitudinal movements, local disturbances, such as pumping ballast, and wet and dry condition influences. Extensive testing presented in reference [10] has shown that it is possible to describe a probability density of the resistance for a given type of track and level of consolidation. Some of the distributions found in the field tests approximate a normal distribution, but it is not necessary to assume such normalcy in the model.

Figure 3-11 presents an assumed resistance distribution. The resistance can be determined at a number of locations using test fixtures and measurements, such as STPT, as discussed in Appendix B.

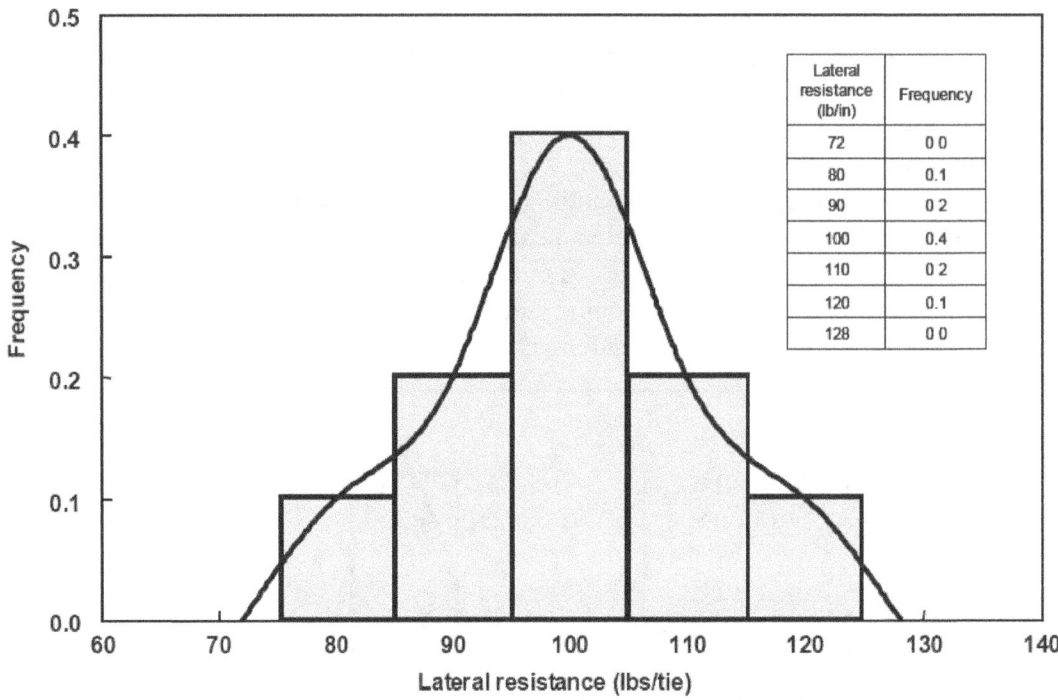

Lateral resistance (lb/in)	Frequency
72	0 0
80	0.1
90	0 2
100	0.4
110	0 2
120	0.1
128	0 0

Figure 3-11. Assumed Lateral Resistance Distribution

Lateral Misalignment Distribution

The allowable lateral misalignment depends on the classification of track in accordance with FRA Track Safety Standards definitions. The current allowable misalignment amplitudes for U.S. track Classes 4–8 are usually given as maximum deviations from the ideal shape over a given chord length. For tangent track, based on a 62-chord length, the permissible deviations are 1.5 for Class 4, 0.75 in for Classes 5 and 6, and 0.5 in for high-speed track Classes 7, and 8.

Track geometry records can be used to evaluate the probability distributions of the misalignment amplitudes. Although the respective wavelengths are also important, their independent distributions are not required as direct inputs because they are automatically generated within the program. Figure 3-12 shows a typical Class 4 alignment distribution. In the figure, the negative values are not present because the example is for a curved track.

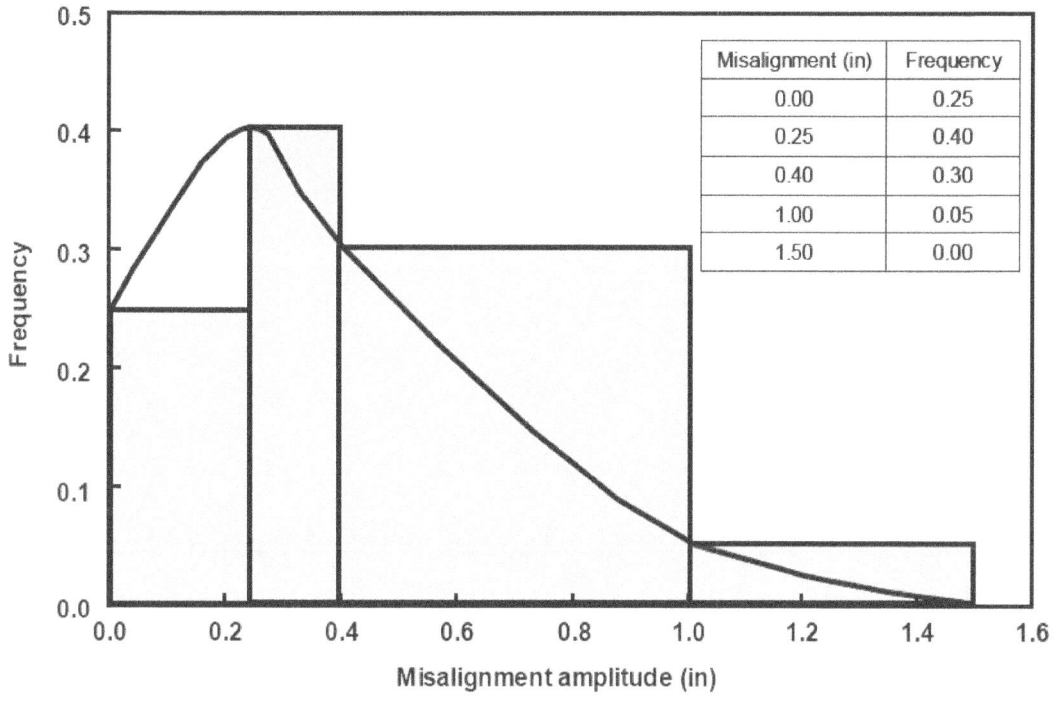

Misalignment (in)	Frequency
0.00	0.25
0.25	0.40
0.40	0.30
1.00	0.05
1.50	0.00

Figure 3-12. Assumed Misalignment Amplitude Distribution for Class 4 Track

For tangent tracks, the negative values are typically folded over to the positive sense so that the largest amplitudes dominate regardless of direction. Table 3-10 shows the deterministic input parameters that are required in the analysis.

Table 3-10. Deterministic Input Parameters Required for CWR-RISK Analysis

Data Inputs for CWR-RISK Illustrative Example	
Rail size (lb/yd):	AREA 136
Tie type:	Wood
Tie weight (lb):	200
Tie spacing (in):	20
Track curvature (deg):	5
Ballast type:	Granite
Tie ballast friction coefficient:	1.2
Torsional resistance (in-kips/rad/in):	120
Longitudinal stiffness (psi):	200
Foundation modulus (psi):	6,000
Vehicle type:	Hopper

<u>Illustrative Example Results</u>

For the illustrative example, the report used the distributions in Figures 3-10 to 3-12 and the parameters in Table 3-10. The output of CWR-RISK is a graph displaying the probability of track buckling as a function of rail temperature as shown Figure 3-13. The figure shows that the probability of buckling is zero (or very close to zero) until the rail temperature reaches a critical temperature denoted by T_c equaling 104 °F. The critical temperature, T_c, is based on a 1 in 1,000 occurring event, taken as 10^{-3} probability value. Beyond this point, the probability increases at a slow rate until the knee in the curve, after which, a steep increase with rail temperature occurs.

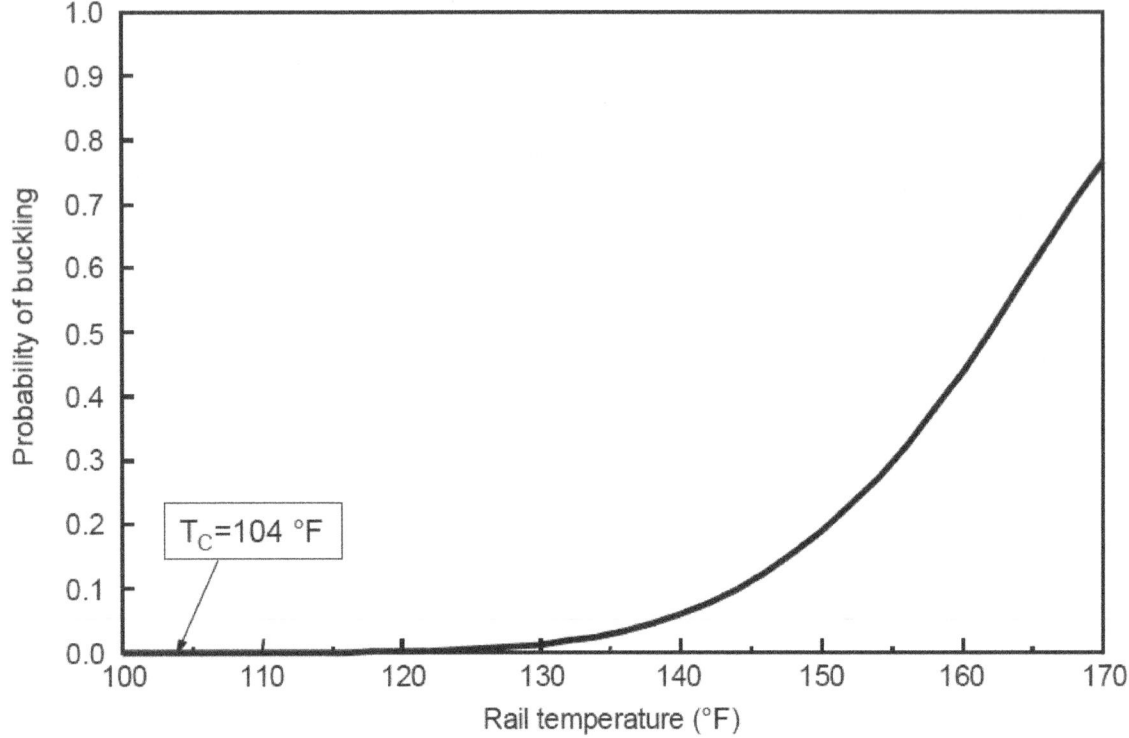

Figure 3-13. Probability of Buckling versus Rail Temperature

As shown in this example, the probability of buckling for a given set of input parameters can be determined as a function of the rail temperature. At low rail temperatures, the probability is zero or very low, and at high temperatures, it can reach 100 percent. With this type of information, several practical issues can be addressed including the following:

- Determination of critical temperature at which slow orders should be imposed
- Estimation of annual number of expected buckles over a given territory for a given annual rail temperature distribution
- Risk information to rail operators on train speeds at temperatures above the critical, depending on the level risk accepted

Chapter 5 will discuss these aspects further, and Appendix C provides the probabilistic theory for CWR-RISK.

3.4   Summary

CWR-SAFE a comprehensive CWR buckling program includes three basic modules for safety analyses.

The CWR-BUCKLE module is a fundamental buckling analysis program using a deterministic approach. The program has been validated by several full-scale tests and requires scientific input parameters. The output includes the buckling response curve, critical temperatures, safe allowable temperatures, energy required to buckle the track, margin of safety, and methods to increase the margin of safety if it is inadequate.

The CWR-INDY module is specially tailored for use by the rail industry. This program uses simple track inputs, such as ballast type, shoulder and crib level, and tie type. Like CWR-BUCKLE, this program provides deterministic values for the buckling temperature, safe allowable temperature, and margin of safety.

CWR-RISK is the third module in CWR-SAFE for which three key inputs, lateral resistance, misalignment, and neutral temperature, require statistical descriptors. All other parameters remain deterministic. The output of the CWR-RISK program is the probability of buckling versus rail temperature.

4. Track Buckling Parameters and Sensitivity Study

This chapter presents a detailed parametric and sensitivity study on the basis of the CWR-BUCKLE program in accordance with the theory presented in Chapter 2 and the program description in Chapter 3. This chapter describes the effects of track resistance, misalignments, curvature, and vehicle parameters on track buckling strength, specifically in terms of influences on $\Delta T_{Bmax/min}$. Chapter 4 will also address special cases representing important aspects and conditions, such as static versus dynamic buckling, weak versus strong lateral strength tracks, and wood versus concrete tie track buckling. Appendix B presents the techniques to measure the key parameters.

4.1 Parametric Study Basis

In most of the parametric sensitivity studies presented below, each parameter of interest is varied through a reasonable range while the remaining parameters are kept at a nominal value in accordance with Table 4-1, unless otherwise specified.

Table 4-1. Ranges and Nominal Values of Buckling Parameters

Parameter		Variable	Range	Nominal Value
Rail Size		Area and moments of inertia	100#RE–141#RE	136#RE
Track resistance	**Lateral**	Lateral resistance peak (F_P)	50–300 lb/in	100 lb/in
		Tie/ballast friction coefficient (μ_f)	0.65–2.0	1.2 (wood) 0.86 (concrete)
	Torsional	Torsional stiffness (τ_o)/fastener	100–5,000 in-kips/rad	1,200 (wood) 500 (concrete)
	Longitudinal	Longitudinal stiffness (k_f)	25–500 lb/in/in	200 lb/in/in
Curvature		Degree of curvature	Tangent - 10 deg	5 deg
Misalignments		Misalignment amplitude (δ_o)	0.5–3.0 in	1.5 in
		Misalignment wavelength ($2L_0$)	L_0=100–500 in	L_0 = 180 in
Track foundation modulus		Vertical stiffness (k_V /track)	2,000–10,000 psi	6,000 psi
Vehicle parameters		Axle load	15,000–75,000 lb	66,000 lb
		Truck center spacing (TCS)	350–700 in	506 in

4.2 Static versus Dynamic Buckling

Chapters 2 and 3 describe the important aspects of static versus dynamic buckling and this section will provide a numerical example to illustrate the key differences. Recalling that static buckling refers to a purely thermal load induced buckling (i.e., without the influence of vehicle loads, dynamic uplift, and train energy inputs), Figure 4-1 shows the buckling response curve for the input parameters noted in the figure.

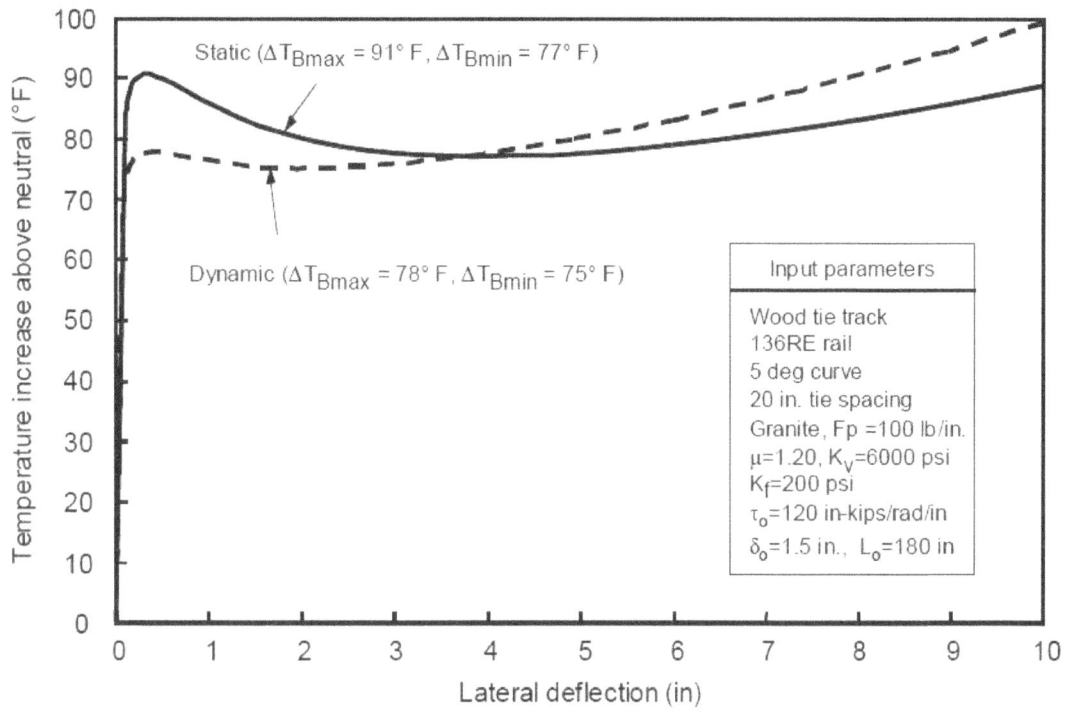

Figure 4-1. Static versus Dynamic Buckling Behavior

The track is expected to buckle out statically at its ΔT_{Bmax}= 91 °F since the track is in a state of unstable equilibriums. The corresponding dynamic case is also shown in the figure, indicating a small buckling regime between ΔT_{Bmax} and ΔT_{Bmin} of 78 °F and 75 °F. From a safety point of view (and in accordance with the safety criterion in Chapter 2), the permissible temperature increase value should be below ΔT_{Bmin} value of 75 °F. As this example illustrates, static considerations provide unconservative numbers hence inadmissible for buckling safety evaluations, and therefore dynamic theory applications are required. Consequently, the use of the dynamic theory is implicit in all the subsequent parametric studies presented here.

Influence of Rail Size

The pertinent rail properties include rail cross-sectional area; cross-sectional moments of inertia for bending in both the vertical (I_{yy}) and horizontal (I_{zz}) planes; and the track self-weight, which includes the weight of rails, ties, and fasteners. Table 4-2 lists these parameters according to the rail size (lb/yd). Figure 4-2 shows the results for the effects of rail size on dynamic buckling for rail sizes ranging from 100 to 140 lb/yd.

Table 4-2. Typical Rail Properties

Rail (lb/yd)	Area (in^2)	Track weight (lb/in)	I_{yy} (in^4)	I_{zz} (in^4)
100	9.95	18.14	49.0	9.4
115	11.25	18.87	65.6	10.8
132	12.95	19.84	88.2	14.6
136	13.35	20.06	94.9	14.7
140	13.80	20.31	96.8	14.8

Properties shown are for a single rail, except for track weight, which includes two rails, wood ties, and tie plate/cut spike fasteners.

The results show that the ΔT_{Bmax} and ΔT_{Bmin} critical temperatures decrease with increasing rail size, with ΔT_{Bmax} showing the larger decrease. Although the rail bending moment of inertia increases with increasing size, the rail cross-sectional area also increases. The increase in area increases the thermal force, which offsets the effect of corresponding increase in bending stiffness, thus reducing the overall buckling strength.

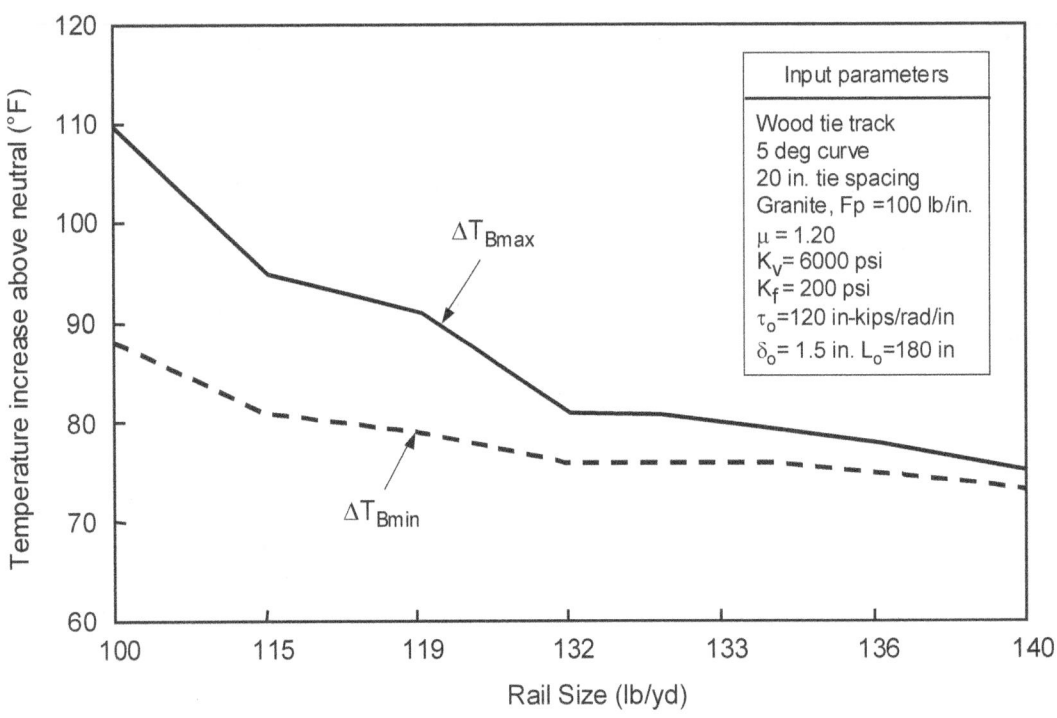

Figure 4-2. Influence of Rail Size

Although a smaller rail size improves buckling strength, it would not be a preferred way to enhance buckling safety since rail size requirements are usually dictated by wheel loads, rail maximum stress/deflection, and fatigue life considerations. Additionally, within the typical heavy-haul rail size range of 132 to 141 lb, the realized buckling strength benefits of rail size are not significant.

56

Influence of Lateral Resistance

The tendency of the track to buckle laterally due to thermal loads is resisted by the ballast lateral reaction forces exerted through the ties. Extensive experimental studies have been performed under FRA sponsorship to characterize this lateral resistance through the use of an STPT method [10, 11, 22], in which a lateral load is applied to an unfastened tie, and the resulting force versus tie displacements are measured, as discussed in Appendix B. Initial force characteristic, Figure 2-8, is approximately linear up to a maximum or peak value (F_P) at a small displacement (usually on the order of 0.25–0.5 in), which is followed by a spring softening effect (tie/ballast friction breaking away) down to a limiting resistance value (F_L), occurring at a limiting or constant lateral deflection denoted by w_L.

Several idealizations of this lateral resistance characteristic have been investigated for computer simulations, such as constant, softening, and full nonlinear (see Figure 4-3), and their influence on $\Delta T_{Bmax/min}$ are available in [4], where it is also shown that the differences between the softening and full nonlinear representations are generally negligible. Although the full nonlinear function is a more complete description of typical lateral response characteristics, it is also mathematically complex, requiring large numbers of iterations for convergence in the computer program. The softening lateral resistance produces results quite close to that of the full nonlinear characteristic; hence, for calculation of buckling response, the idealized softening lateral resistance characteristic is chosen, Figure 4-3 B.

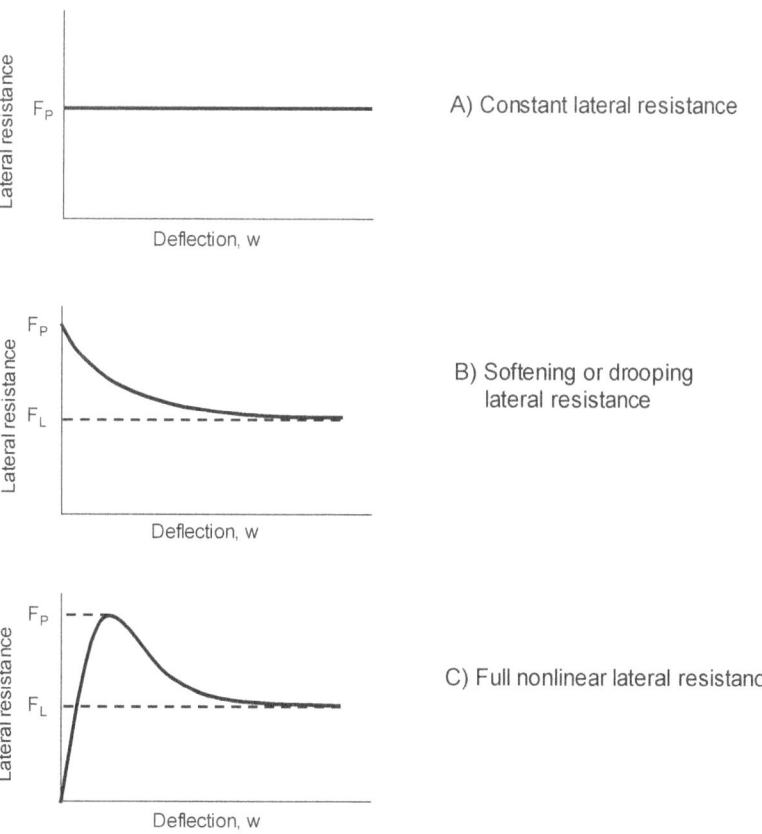

Figure 4-3. Lateral Resistance Idealizations

57

The softening resistance function is defined by two variables: the ratio of the limiting and peak resistances and the deflection at the limiting resistance. These parameters depend on the type of ballast and level of ballast consolidation. To simplify and reduce the number of parameters required for definition of the softening resistance, a series of tests were conducted [9, 10] to correlate F_L and w_L to F_P, which is a parameter that can be directly measured using single tie push tests. On the basis of this experimental work, approximate linear relationships were developed between F_P and F_L and between F_P and w_L as shown in [9]. Table 4-3 lists these correlations for granite and slag ballast. Thus, the peak lateral resistance, F_P, is the primary parameter that is used to study the effects of lateral resistance on buckling strength. Should the requirement exist to use F_P, F_L, and w_L in a specific analysis, CWR-BUCKLE has the optional feature to handle that case.

Table 4-3. Correlations for "Softening" Lateral Resistance

Ballast type	F_P/F_L correlation	F_P/w_L correlation
Granite	$F_L = 0.3\ F_P + 25$	$w_L = 0.025\ F_P + 2.6$
Slag	$F_L = 0.06\ F_P + 30$	$w_L = 0.009\ F_P + 3.5$

Wood tie track; units for F_P and F_L are lb/in; and units for w_L are inches.

Influence of Peak Resistance (F_P) on Buckling

To determine the effects of lateral resistance on buckling, the parameter F_P is varied over a practical range representing very weak or recently maintained track ($F_P = 50$ lb/in) up to very strong, well consolidated track ($F_P = 300$ lb/in). Figure 4-4 shows the buckling temperature results. The results show that for very weak tracks, progressive buckling can occur for temperature increases of 50 °F or less for the parameters considered. Such tracks are vulnerable to buckling in summer, even with a rail neutral temperature of 80 °F.

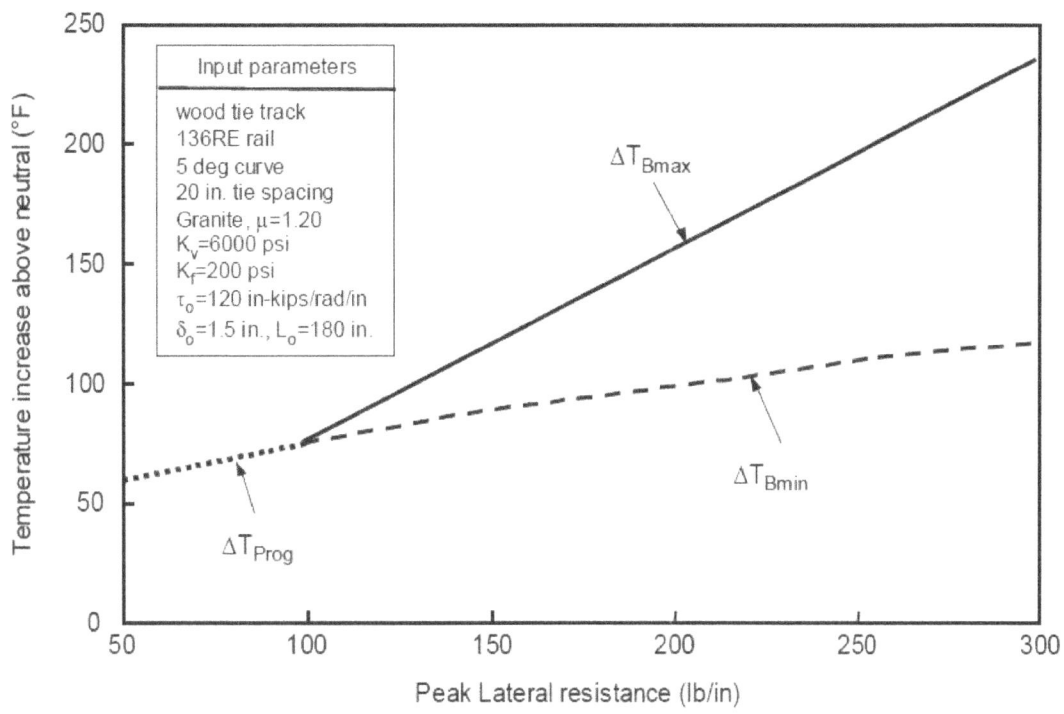

Figure 4-4. Influence of Peak Lateral Resistance, F_P

As F_P increases, the upper and lower buckling temperatures increase. For very strong tracks (F_P >200 lb/in), the minimum buckling temperature increase value is quite high (100 °F) even for the 5° curve used in the example.

Clearly, it is important to maintain a high lateral resistance to reduce the track buckling potential. The railroads can use the lateral resistance condition influencing parameters (i.e., consolidation level, shoulder width, cribs content, tie type) to help keep F_P high, thereby better ensuring CWR buckling safety.

Influence of Limiting Resistance (F_L)

The nonlinear lateral resistance characteristic (i.e., F_P versus F_L behavior) has a very important influence on the buckling response, as shown in Figure 4-5, with the key feature of its impact on ΔT_{Bmin} (and very little on ΔT_{Bmax}). As can be seen, as F_L decreases from a value of 3600 lb/tie to1400 lb/tie (while keeping F_P constant at 3600 lb/tie), the ΔT_{Bmin} value decreases from 121 °F to 89 °F. This is very important because the ΔT_{Bmin} value mostly controls the T_{all} determination in the buckling safety criterion as discussed in Section 2.3.2.

59

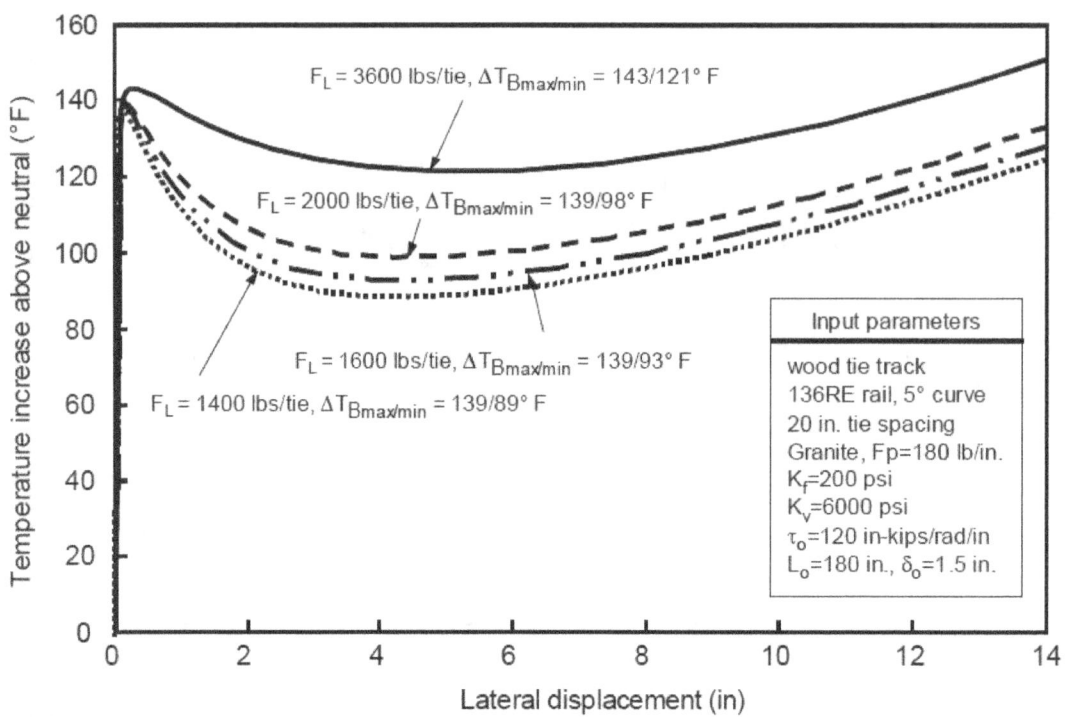

Figure 4-5. Influence of Limiting Resistance

Strong versus Weak Resistance Tracks

To provide an overall influence assessment of lateral resistance on the track buckling response, a comparison is made between lateral resistance conditions that would represent weak, average and strong lateral resistances in accordance with the representations shown in the right hand side of Figure 4-6.

For the baseline parameters considered, the strong track exhibits high and distinct ΔT_{Bmax} and ΔT_{Bmin} values of 140 °F and 100 °F, respectively. The average track shows a lower and almost identical ΔT_{Bmax} and ΔT_{Bmin} values of 96 °F and 93 °F, while the weak case exhibits a highly progressive behavior. From the buckling safety considerations of Chapter 2, the strong track has a buckling regime of between 100 °F and 140 °F, with the safe temperature being 100 °F above neutral. Similarly, for the average condition, the safe temperature must be below the 93 °F value. For the weak track's progressive case, beyond 70 °F large displacements occur, so the safe temperature should be below this number.

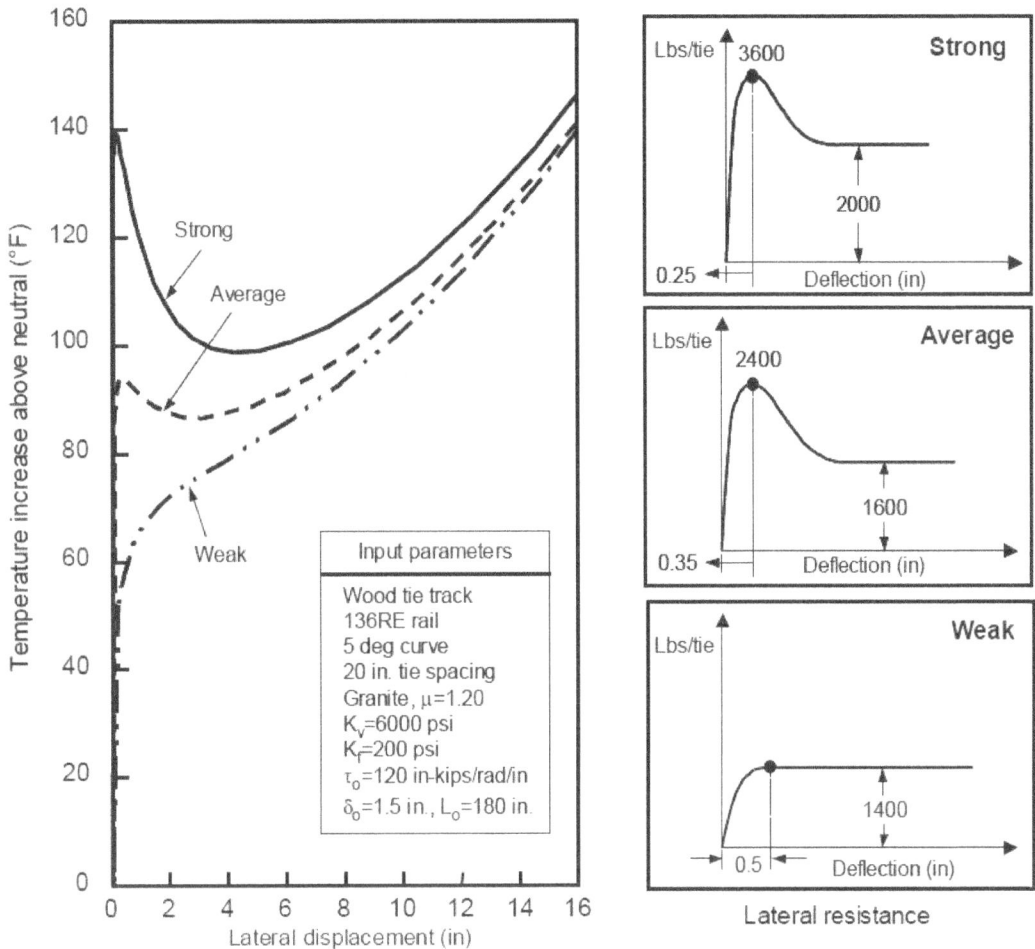

Figure 4-6. Buckling Response Behavior for Weak, Average, and Strong Tracks

Influence of Tie/Ballast Friction

The tie bottom surface roughness is an important parameter as it determines the component of the base resistance. If all other conditions remain unchanged (i.e., rail size, tie weight, ballast type), the lateral resistance should increase with tie bottom roughness; hence the buckling strength should also increase. In the case of wood ties, the tie bottom roughness controls the bottom surface friction resistance as the ballast tends to lock itself into the tie bottom. The concrete ties are relatively smooth surfaced, and even when the tie bottoms and sides are artificially roughened at the time of manufacture (as in some scalloped tie designs), the surfaces tend to become smooth in course of service life due to the pounding and grinding action between the tie and the ballast particles.

The tie bottom roughness factor is artificially expressed as a friction coefficient which is defined as the ratio of the measured base resistance to tie self-weight. Although in principle it is similar, this coefficient should not be confused with the Coulomb friction between two surfaces, whose value never exceeds unity. The lateral resistance can be expressed as the sum of the base, side, and end shoulder resistance components of the tie:

$$F = F_b + F_s + F_e \qquad\qquad\qquad \text{Equation } (4-1)$$

61

Defining the coefficient μ_f in terms of weight of tie, Q (including the weight of rail and fasteners):

$$F_b = \mu_f Q \hspace{6cm} \text{(Equation 4 – 2)}$$

μ_f can be considered an index of tie bottom roughness. If the vehicle contributes an additional load of R_v (lb/in.) on the tie, the dynamic resistance can be calculated from Equation 19 in Appendix A. The coefficient μ_f in Equation 19 is strictly a function of total tie reaction. The average value of μ_f typically varies in the range from 0.65 to 2, and Appendix B discusses its determination. For this parametric exercise, Table 4-4 shows assumed values of the three resistance components for wood tie track with a typical weight Q = 20 lb/in.

Table 4-4. Three Components of Resistance and Their Assumed Variations with μ_f

μ_f	F_s (lb/in)	F_e (lb/in)	F_b (lb/in)	F_p (lb/in)
2.0	48	24	48	120
1.0	48	24	24	96
0.65	48	24	16	88

Figure 4-7 shows the effects of the tie-ballast friction coefficient. As expected, an increase in tie bottom surface roughness (increasing μ_f) increases both buckling temperatures. Over the range studied here, the increase of the buckling temperatures is less than 10 °F.

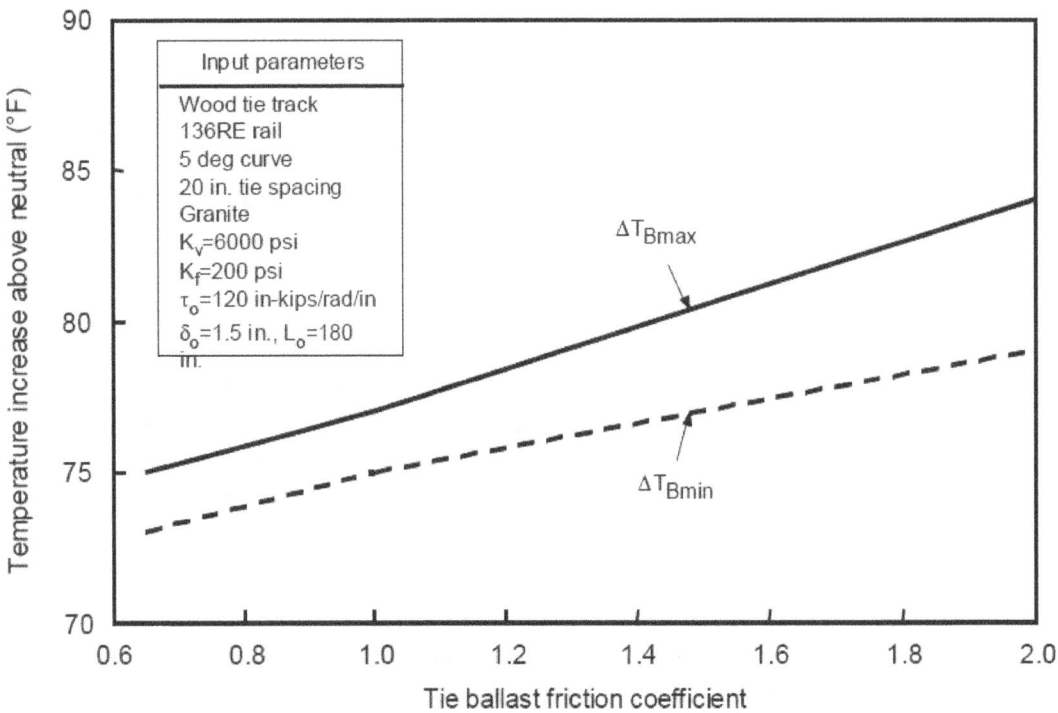

Figure 4-7. Influence of Tie-Ballast "Friction" Coefficient

Care must be exercised in attempting to evaluate μ_f influence without making the direct adjustments in F_P along the lines of Table 4-4 (i.e., if one assumes a fixed F_P and runs the variations on μ_f, results opposite in trend to Figure 4-7 will be obtained).

Influence of Torsional Resistance

As noted in Chapter 2, torsional resistance is exerted on the rails by the fasteners, and the resistance characteristics vary with tie and fastener types and influences the rigidity of the track structure in the lateral plane, hence its buckling strength. Based on torsional resistance measurements from tests conducted for various tie and fastener types, the response can be idealized as a linear response as a function of rotation:

$$\tau = \tau_o \theta \qquad \text{(Equation 4-3)}$$

where

$\quad \tau$ = applied torque per fastener
$\quad \tau_o$ = torsional stiffness per rail seat
$\quad \theta$ = rotation angle

Table 4-5 gives typical ranges of stiffness values for single rail/fastener obtained from test data presented in [9]. Consistent with the model's track-beam assumption the values of τ_o in Table 4-5 are to be multiplied by 2 for the two rails and then divided by the tie spacing to obtain the correct track/unit length dimension. The effects of fastener torsional resistance on buckling are examined using the linear torsional stiffness values ranging from 0 to 3500 in-kips/rad. Figure 4-8 and Figure 4-9 show the results.

Table 4-5. Torsional Stiffness Values

Type of Tie	Fastener	τ_o (in-kips/rad)
Hardwood	Pandrol	3,700–7,400
Hardwood	Cut Spikes (4)	800–1,400
Concrete	Pandrol	120–520
Concrete	McKay	300–440

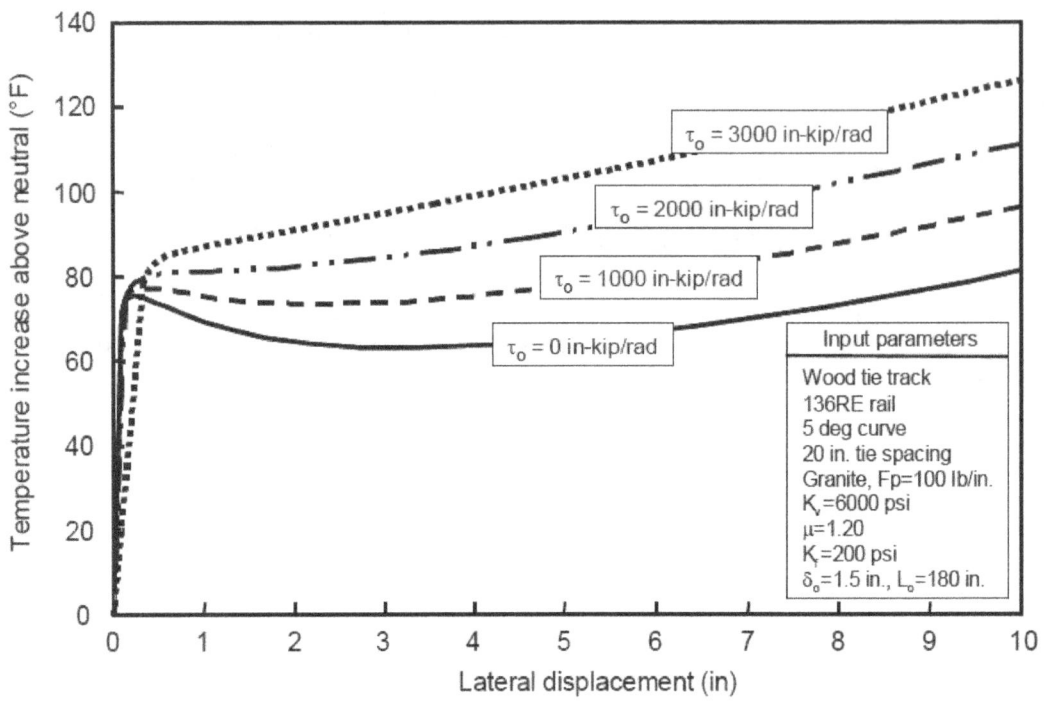

Figure 4-8. Torsional Stiffness Influence on Buckling Response

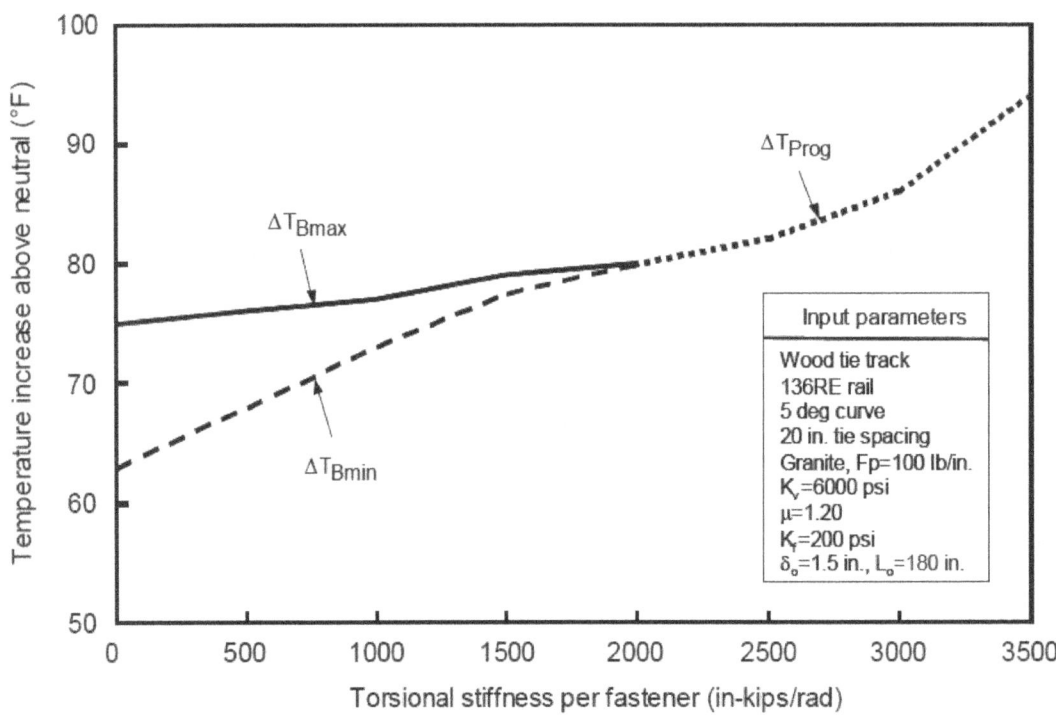

Figure 4-9. Influence of Torsional Stiffness on Buckling Temperatures

As can be seen from Figure 4-9, both $\Delta T_{Bmax/min}$ values increase with increasing torsional stiffness, and in fact, the response curves change from nonprogressive to a progressive (which is contrary to other parametric influences where typically a weakening of a parameter results in a

64

progressive buckling response). The fact that the lower buckling temperature, ΔT_{Bmin}, is the more sensitive to torsional resistance changes (as illustrated in Figure 4-9) becomes important in safety considerations since it is the ΔT_{Bmin} value that controls the buckling safety criterion. The results further indicate that the better the rail/tie fastener connections are, (i.e., the higher the fastener's torsional resistance) the higher is the buckling strength.

Effects of Longitudinal Resistance

Track longitudinal resistance is the resistance offered by the fasteners to the rails and by the ballast to the ties against longitudinal motion. In CWR, it becomes very important in providing the constraint against thermal expansion, hence in the generation of longitudinal forces and neutral temperature constancy. In principle, the two components of resistance can be considered as two springs in series, so the net resistance is controlled by the weaker of the two components. Therefore, in good strong ballast, the fastener resistance dominates the longitudinal restraint, so when ties are ineffectively anchored the longitudinal resistance can be very low. Conversely, with properly anchored ties, the resistance offered by the ballast becomes important. Data from tests (Appendix C) conducted at TTC indicate that longitudinal resistance may be idealized as in Figure 4-10. Due to the small longitudinal displacement that occurs during buckling (usually less than 0.25 in as evidenced in several buckling tests), a linear characteristic can be assumed in the analysis so that:

$$f = k_f u \qquad\qquad\text{(Equation 4-4)}$$

where

k_f = longitudinal stiffness
u = longitudinal displacement

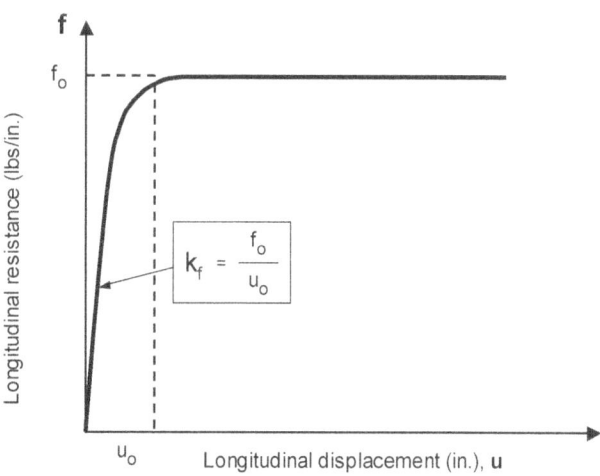

Figure 4-10. Typical Longitudinal Resistance Characteristic

Table 4-6 lists typical measured values of k_f from tests [9] for various tie and ballast conditions for wood ties. For this parametric study, a typical range of stiffness values (25 to 500 psi) is used. Figure 4-11 shows the buckling results. As can be seen, the lower critical temperature is more sensitive to the changes in longitudinal stiffness, and the upper critical temperature is essentially independent of the changing stiffness. Since k_f influence on T_{Bmin} is relatively small, the need to determine or know k_f exactly is not as important as some of the other parameters.

65

However, track longitudinal stiffness is very important in controlling neutral temperature variations and, as such, plays an important role in buckling safety management.

Table 4-6. *Typical Measured Longitudinal Stiffness Values*

Tie condition	Ballast condition	k_f (psi)
ETA	Consolidated	324
EOTA	Consolidated	254
ETA	Tamped	213
EOTA	Consolidated ½ crib	126
E3TA	Tamped	178

ETA = every tie anchored
EOTA = every other tie anchored
E3TA = every third tie anchored

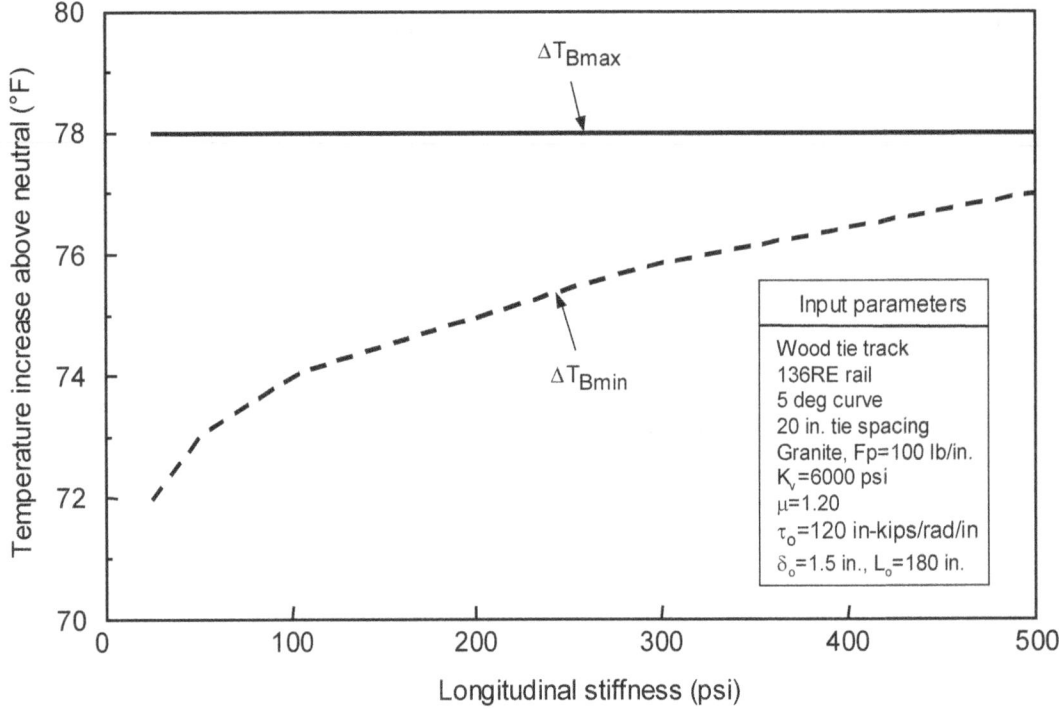

Figure 4-11. Influence of Longitudinal Resistance

Influence of Track Curvature

The effects of track curvature are investigated using curves ranging from tangent to 9°. To illustrate the influence of the lateral resistance on curved track buckling, the buckling temperatures are calculated for the three cases of strong, average, weak resistances of Section 4.3, representing tracks of varying peak lateral resistances of F_P = 70, 120, and 180 lb/in. Figure 4-12 shows the results of this study. In each case, the results show that increasing curvature reduces the upper and lower buckling temperatures. For high curvature tracks or tracks with a

66

weak lateral resistance, the buckling temperatures are drastically reduced in comparison to tangent track. It is important that, for weak tracks (such as with F_P = 70 lb/in.), progressive buckling can occur in curves of about 7° and higher. The safe temperature increase for such tracks thus falls below 55 °F and can be vulnerable to buckling in summer, even with a rail neutral temperature of 80 °F. Figure 4-13 portrays the curvature influence on $\Delta T_{Bmax/min}$ specifically, showing the decreasing buckling strength influence with increasing track curvature.

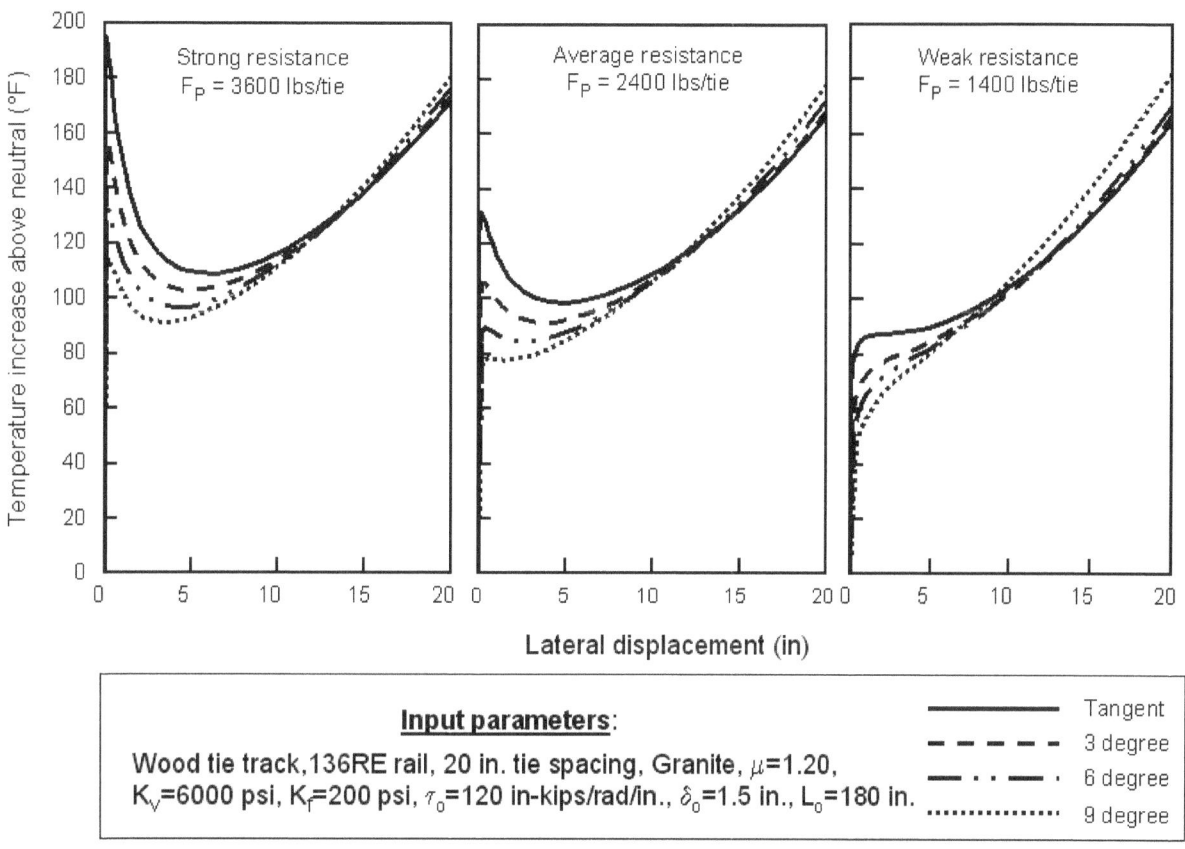

Figure 4-12. Effect of Curvature on the Buckling Response Behavior

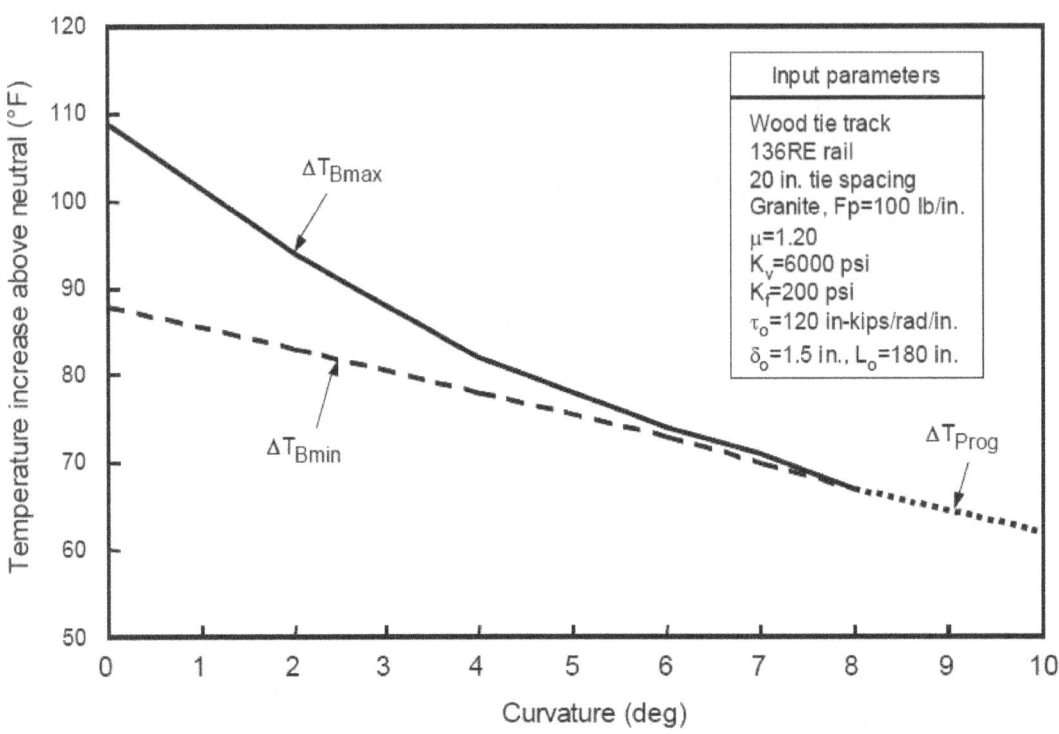

Figure 4-13. Influence of Curvature on Buckling Temperatures

Effects of Initial Misalignments

Misalignment parameters include the track misalignment amplitude (δ_0) and misalignment wavelength ($2L_o$). Several functional forms are available for idealization of the misalignment shape, including sinusoidal, parabolic, or fourth order polynomial representations. Of these, the fourth order shape is employed as it is more consistent with the Shape I buckling mode shape assumption, which requires zero end slopes and deflection.

Influence of Misalignment Amplitude

The misalignment effects are examined first by using typical misalignment amplitudes ranging from 0.5 to 3 in, representing FRA Class 8 through Class 2 alignment defects. In this example, the corresponding misalignment half wavelength L_0 is fixed at 180 in for all amplitudes. Figure 4-14 shows the results. It should be noted, however, that in reality a specific misalignment amplitude has its own corresponding wavelength, consistent with the track-beam's bending flexibility. In fact, the model has an internal wavelength computational algorithm which can automatically calculate a wavelength for specific amplitude. As Figure 4-14 indicates, both buckling temperatures decrease as the misalignment amplitude increases, with the upper critical temperature being more sensitive to these changes. The changes in the lower critical temperature are more modest, and progressive buckling occurs for misalignment amplitude of about 2 in.

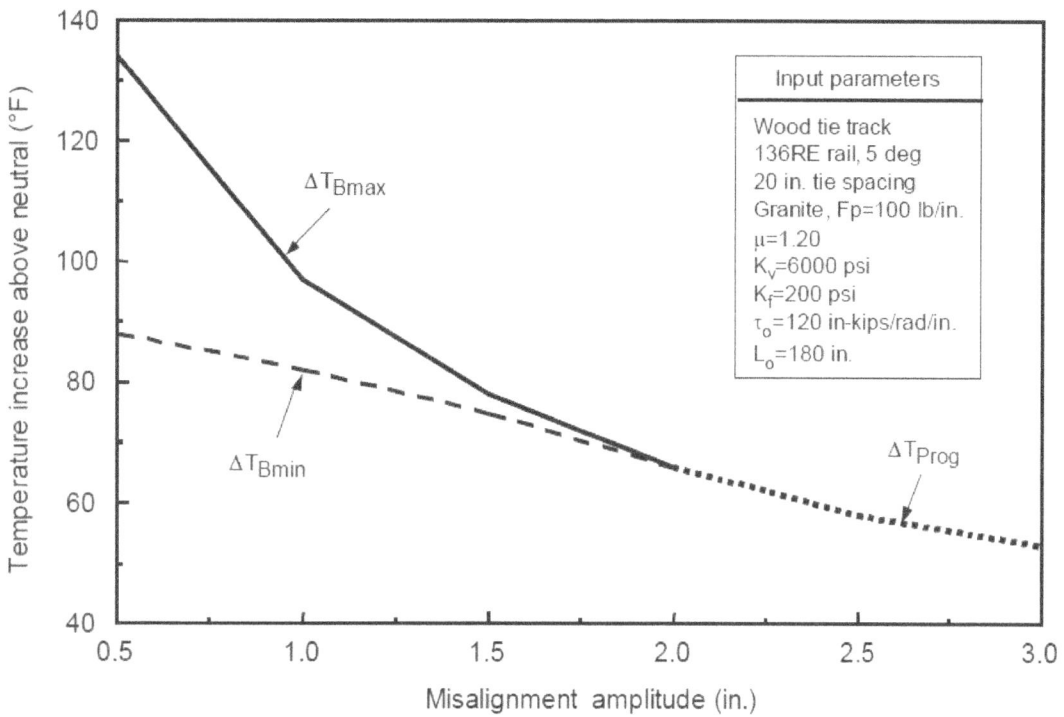

Figure 4-14. Influence of Misalignment Amplitude on Buckling Temperatures

Influence of Misalignment Wavelength

The misalignment wavelength effects are studied using fixed misalignment amplitude of 1.5 in and varying misalignment half-wavelengths from 150 to 500 in. Figure 4-15 shows the results. The lower critical temperature is relatively insensitive to the effects of wavelength; however, the upper critical temperature decreases sharply as the wavelength is reduced. Progressive buckling condition is reached at a half wavelength of approximately 150 in, when the ratio of misalignment amplitude to misalignment wavelength is largest.

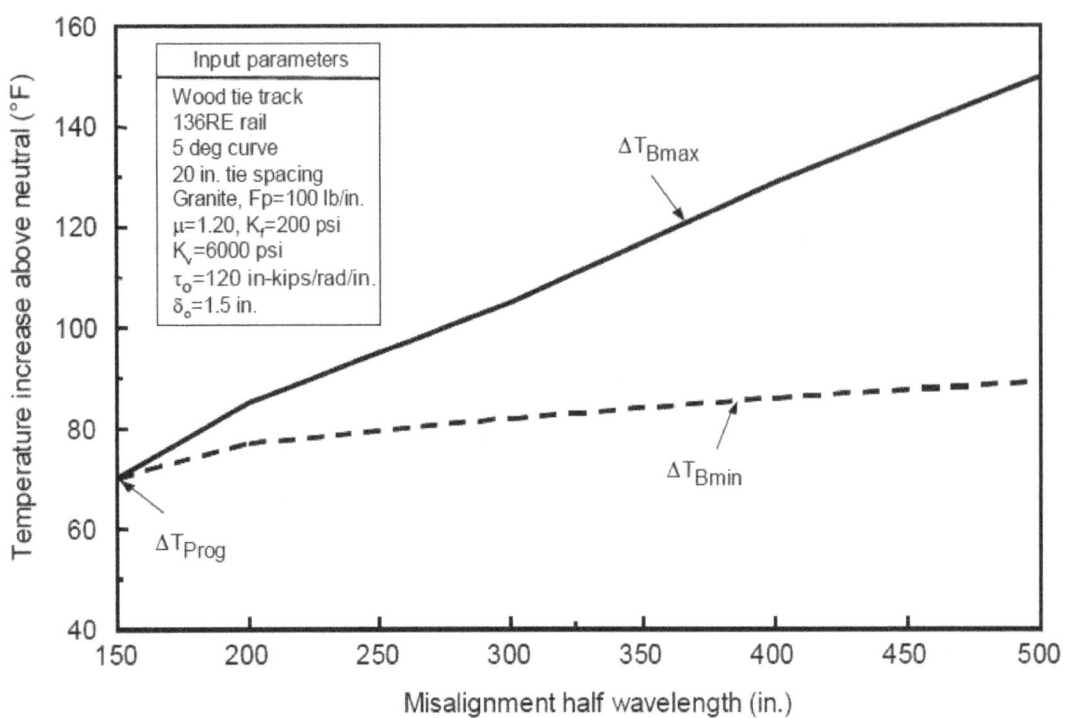

Figure 4-15. Influence of Misalignment Wavelength on Buckling Temperatures

The foregoing results indicate that control of line defects in terms of amplitudes and wavelengths helps in increasing the CWR track buckling strength. Sharp kinks with small wavelengths should be minimized in revenue service conditions.

Effects of Track Foundation Vertical Stiffness

As noted in Section 2, the presence of vehicle loads causes an uplift in the track, which is partially dependent upon the track foundation stiffness. The effects of foundation stiffness are examined using a typical range of 3,000 to 10,000 psi and are shown in Figure 4-16. ΔT_{Bmax} is more sensitive to stiffness variations than ΔT_{Bmin}, and for stiffness values below 4,000 psi the response becomes progressive.

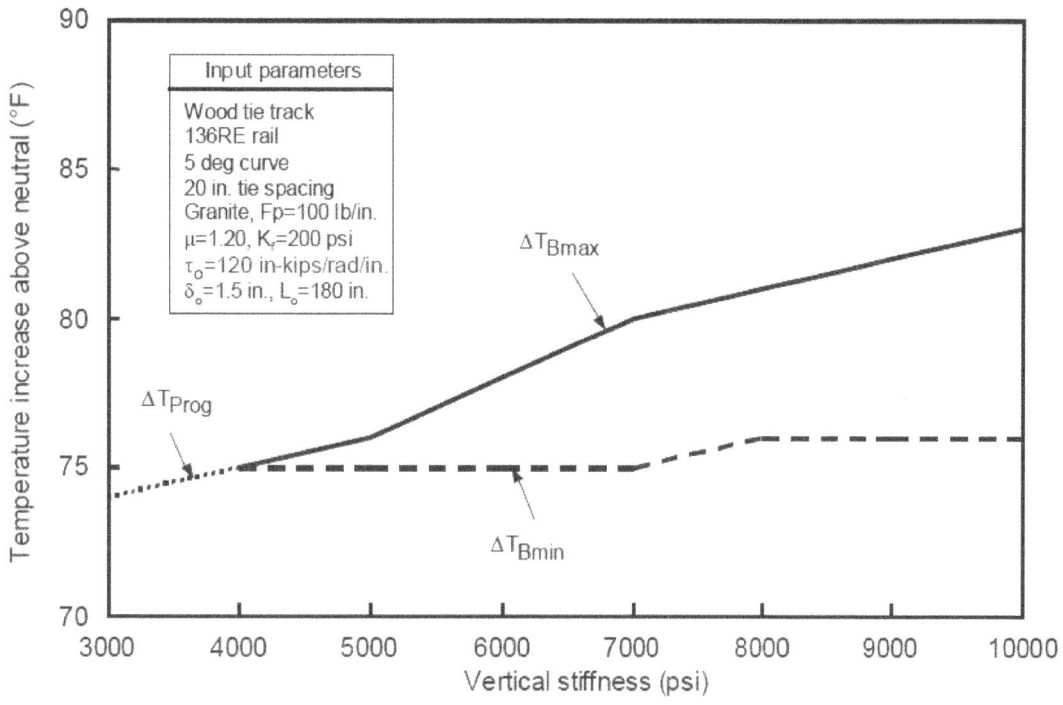

Figure 4-16. Influence of Track Foundation Vertical Stiffness

Effects of Vehicle Parameters

As discussed in Section 2 and implied by Figure 4-1, the effect of vehicle loads is to cause rail uplift and thus reduction of buckling strength in comparison with the static buckling case. The primary vehicle parameters controlling the uplift are the axle loads and the truck center spacing (TCS), which vary with vehicle size and type.

Influence of Axle Loads

To examine the effects of these vehicle parameters on buckling, axle load are first studied using a typical hopper car dimensions. The axle load is varied to reflect changes in the car's cargo weight, ranging from the empty weight load of 15,575 lb to the maximum gross weight load of 65,750 lb. The axle spacing and truck center spacing for this car are 70 and 506 in, respectively. All other parameters are set at the default values. Figure 4-17 shows the buckling results. The lower critical temperature is relatively insensitive to the effects of axle load and is essentially constant. However, the upper critical temperature decreases with the increasing axle load. This is an important aspect of dynamic buckling since the rapid decrease in the upper critical temperature quickly reduces the energy barrier for buckling. The practical implication is that lighter cars tend to have larger buckling margins of safety than heavier cars.

71

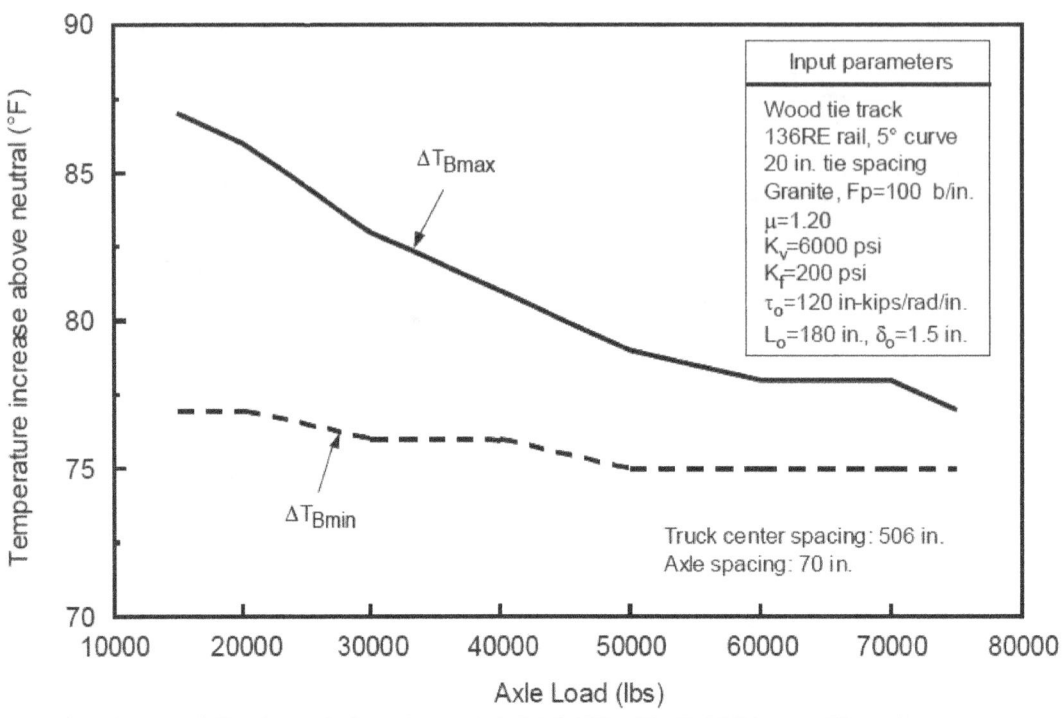

Figure 4-17. Influence of Axle Load on Buckling Temperatures

Effect Truck Center Spacing

The effects of vehicle truck center spacing are examined by varying TCS from 350 to 700 in to reflect changes in car length, while keeping the axle spacing at 70 in and an axle load of 65,000 lb constant. Figure 4-18 shows the results. For small values of TCS (less than 470 in for the example shown), the behavior is progressive and with increasing ΔT_P values. The upper and lower buckling temperatures diverge at a TCS of about 480 in. The upper buckling temperature then increases while the lower buckling temperature continues to decrease with a slight upswing for large TCS values. At these large TCS values, the upper and lower temperatures approach the values predicted by static buckling theory (91 °F and 77 °F, respectively, as shown by Figure 4-1) since the effect of the vertical load is not felt in between the trucks, thereby simulating a static buckling condition. It can be seen that critical truck center spacing is about 480 in corresponding approximately to that of hopper cars. This is considered critical because this has the lowest temperature at which progressive buckling can occur (i.e., no energy barrier as compared to the longer truck center spacing).

72

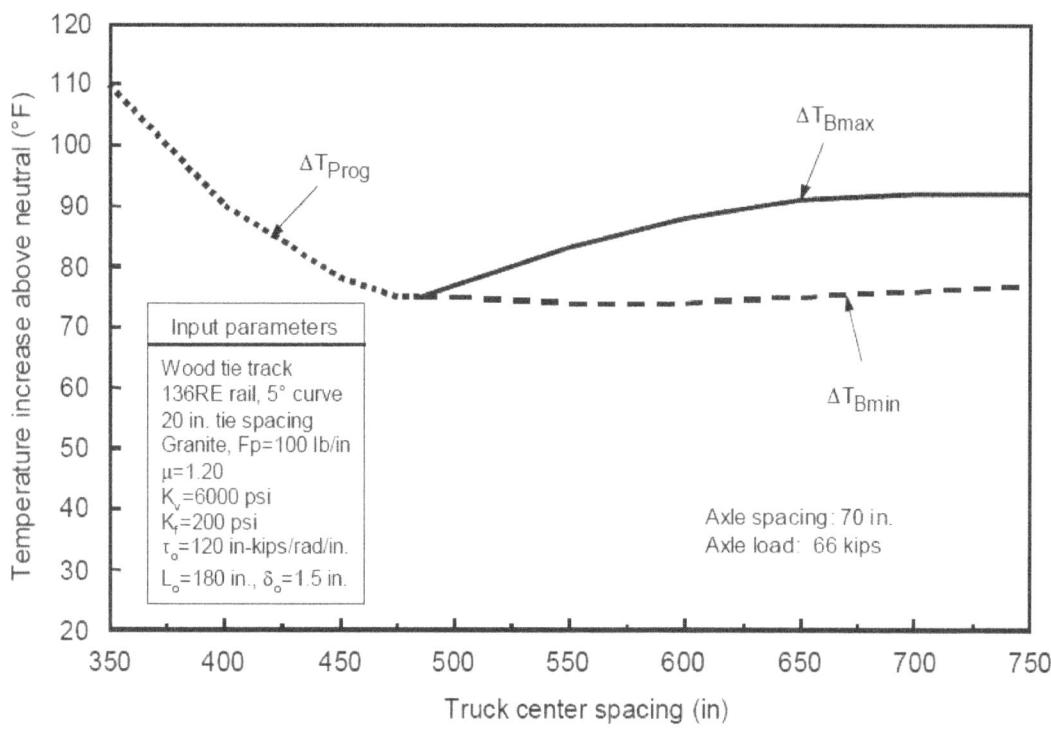

Figure 4-18. Influence of TCS on Buckling Temperatures

Effects of Track Structure: Concrete versus Wood Ties

Concrete tie CWR track is generally regarded as stronger than the wood tie track from a buckling strength point of view. The higher strength of concrete tie track is attributed to its larger tie weight, size and shape, and usually better quality ballast, all contributing to increase the lateral resistance. However, other factors, such as the higher friction coefficient of wood ties, smaller tie spacing for wood ties (20 in for wood versus 24 in for concrete tie track), and typically higher torsional resistances for cut spike fasteners, become important in offsetting the perceived concrete tie weight, size, and shape advantage.

To make a reasonable comparison, in this report we will postulate an average condition for the wood and concrete tie track while keeping as many parameters constant as possible (i.e., rail size, curvature, ballast type, and track alignment condition). The average condition for wood ties is taken as that shown in Figure 4-6 with a peak resistance of 120 lb/in, while the average condition for concrete ties is taken as indicated in Figure 4-19 with a peak resistance of 142 lb/in. as measured in recent tests [26]. As Figure 4-19 indicates, for the average conditions stipulated, the concrete tie track has a higher ΔT_{Bmax} but a lower ΔT_{Bmin} than for wood tie tracks. Because of the larger difference between the concrete tie ΔT_{Bmax} and ΔT_{Bmin} than for the wood tie, the buckling energy at the respective ΔT_{Bmin} values is higher for concrete than wood, indicating that, while buckling safety at the concrete tie ΔT_{Bmin} may be assured, at the wood tie ΔT_{Bmin} it is not. According to the safety criterion of Section 2.3.2, the safe allowable temperature $T_{all\ (wood)}$ is 84 °F while the $T_{all\ (concrete)}$ is 80 °F, implying slightly better buckling strength for wood tie tracks for the parameters chosen. Figure 4-20 shows a similar comparison for a weak or recently maintained track. For the weak wood tie track, the response is progressive buckling, while for the concrete tie, track although distinct $\Delta T_{Bmin/max}$ exist, the response is close to progressive. For this weak condition, the buckling safety criterion gives a $T_{all\ (wood)} = 60\ °F$, and $T_{all\ (concrete)} = 56\ °F$, again showing a slightly higher buckling strength for wood tie tracks. On the other hand, the wood tie track buckles more progressively, therefore construed to be weaker in the buckling sense. At $T_{all\ (wood)} = 60\ °F$ the track has already incurred an additional half inch of lateral deflection adding to the initial misalignment of 1.5 in. It is evident that progressive buckling should be avoided due to rapid incurrence of lateral deformations with temperature increase.

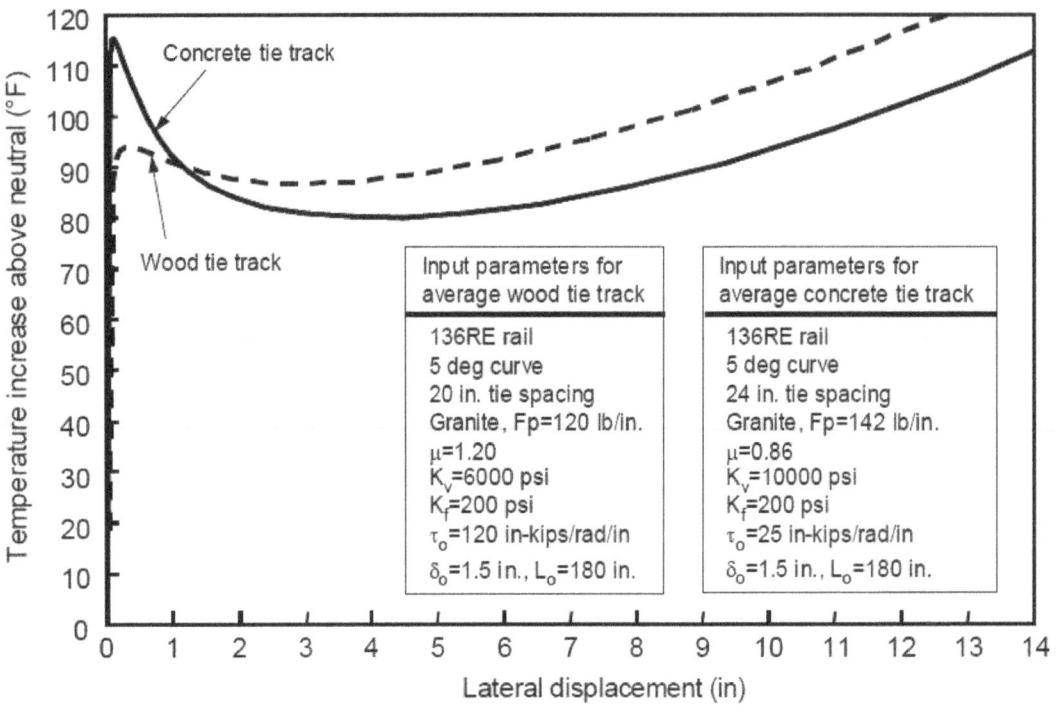

Figure 4-19. Buckling Response for "Average" Wood versus Concrete Tie Tracks

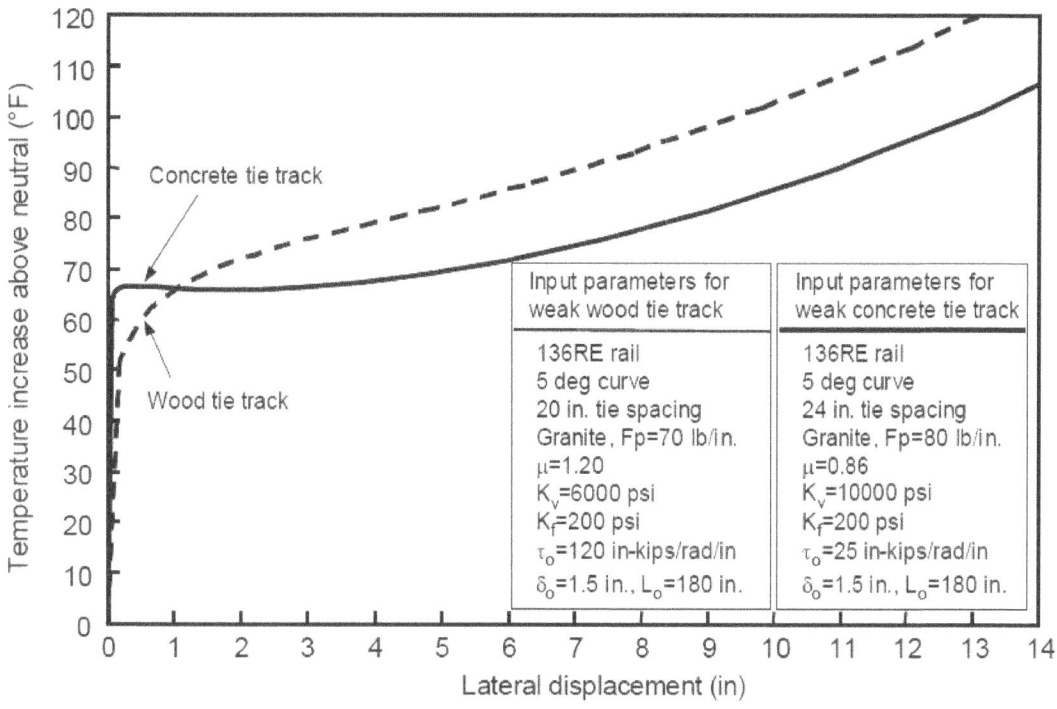

Figure 4-20. Buckling Response for "Weak" Wood versus Concrete Tie Tracks

Summary

The results of the parametric studies show that several parameters have a significant impact on the buckling temperatures, $\Delta T_{Bmax/min}$, notably track lateral and torsional resistances, misalignments, and track curvature. As will be discussed in Chapter 6 on the safety aspects of CWR, for safety considerations, the ΔT_{Bmin} values are the more important ones since they drive the safety criterion. Special cases addressing conditions and features, such as static versus dynamic buckling, weak versus strong lateral resistance tracks, and wood versus concrete tie track buckling, have also been presented with their implications on the track buckling response. It is important to note that the results presented represent trends and sensitivity influences for a set of fixed parameters. For a specific track type or condition, actual representative parameters should be used for buckling analysis. Table 4-7 presents a summary of the parametric analysis results in terms of desirable values for providing high buckling strength.

Table 4-7. Desired Parameters for High Buckling Strength

Parameter	Desired Value	Comments
Rail size	Small	Buckling strength goes up with decreasing rail size; however, rail size selection is based on fatigue and wheel load considerations.
Peak lateral resistance	High	Buckling strength goes up with increasing lateral resistance. After maintenance, ensure adequate consolidation by traffic or mechanical stabilization. Also, ensure/maintain adequate ballast shoulders and crib content.
Tie/ballast friction coefficient	High	Buckling strength increases with increasing tie/ballast friction. Consolidation tends to increase the coefficient. For concrete ties, rough bottom surfaces help.
Torsional resistance	High	Buckling strength increases with fastener torsional resistance. Avoid spike killed ties and broken or missing insulators on concrete ties.
Longitudinal resistance	High	The benefit of this parameter is more in controlling the rail neutral temperature by limiting longitudinal movement. Anchor effectiveness is very important, and ETA versus EOTA is better.
Curvature	Low	Buckling strength decreases with curvature. To counteract this influence, maintain high neutral temperature and lateral resistance.
Misalignment amplitude	Small	Good alignment is very important. More frequent track geometry inspections are useful, as is controlling/monitoring curve breathing.
Misalignment wavelength	Large	
Track foundation modulus	High	Vertically stiffer track is better from buckling point of view.
Axle load	Small	Increased axle loads tend to reduce the upper buckling temperatures but with very little influence on ΔT_{Bmin}. High axle loads also require larger rail sections, which can further reduce buckling strength.
Truck center spacing	Small or Large	Within the range of typical truck center spacing, the intermediate values tend to be the worst.

5.   Probabilistic Evaluations of Track Stability

This chapter will use the probabilistic theory presented in Chapter 3 to define probabilistic measures of track buckling potential, based on the input parameters (lateral resistance, neutral temperature, and misalignment) being expressed in statistical terms. Unlike the case of the deterministic approach, which gives the margin of safety that could vary from location to location; the probabilistic approach can give a gross index of the buckling potential on a given territory. The probability of buckling at a given rail temperature can be expressed by the number of expected buckles over a given track section. From this and knowing the annual rail temperature over the region, one can compute the probable number of buckles in a year. This type of information will be valuable to the operating agencies for the following reasons:

- The industry can compare the predicted annual number of buckles using the probabilistic methodology in a given territory to the actual reported incidents and accidents attributed to track buckling and this information for planning future preventative maintenance programs.
- By parametric studies, one can determine the optimum parameter levels (lateral resistance, neutral temperature, and misalignments) to minimize the annual number of buckles and thus reduce the number of buckling accidents.
- For a given level of buckling probability (risk acceptance), railroads can perform trade-offs between maintaining high resistance, improving lateral alignment conditions or restressing the rails to higher neutral temperatures.
- The probabilistic theory will also be useful in the decision-making process on slow orders when the rail temperature is high. Speed reductions and the temperatures for their imposition can be rationally determined using a risk-based approach and the probabilistic buckling theory.

Annual Number of Buckles

Chapter 3 presents the approach to determine the probability of buckling at a given rail temperature. Since the rail temperature changes daily and seasonally, i.e., producing its own temperature frequency, it becomes important to evaluate the cumulative probability of buckling over a year. This will be called the annual probability of buckling, P_a, from which the expected number of buckles over a chosen length of track section can be determined as explained in the following paragraphs.

To determine P_a, the annual rail temperature (T_R) frequency distribution for the year is required. This can be derived from the past records on the rail temperature or alternately they can be deduced from the air temperature records available from the U.S. Meteorological Society. If the air temperature is used as a basis, the rail temperature must be adjusted higher than the air's since the rail acts as a black body absorbing heat. The difference between the rail and the air temperature can be 25–35 °F based on the field data. Then the annual probability of buckling is the product of the probability of buckling at the rail temperature P_b and the annual frequency of the rail temperature, P_T, or:

$$P_a = P_b P_T \qquad \text{(Equation 5-1)}$$

Chapter 3.3 identifies the critical temperature, T_c, above which the buckling probability becomes significant. Since P_b is zero or very small for $T_R < T_c$, the rail temperature frequency for $T_R < T_c$ is not important. What is important is the number of days in a year for the rail temperature, T_R, to be in the range of T_c and above to the maximum rail temperature, T_M.

Expected Number of Buckles per Annum

To derive the expected number of buckles in a year, one must first define an appropriate track segment length that is vulnerable to a buckling occurrence. For this, a track segment equal to that of one car length is chosen since buckling usually takes place in the uplift region under a car, as discussed in Chapter 2. If one considers that each car is capable of inducing a buckle at the segment, then the annual number of potential buckling events at the segment is the total number of cars passing over the segment in a year. However, if one buckle occurs at this segment under one car in a train, then all potential buckles under all other cars in the train can be ignored since one buckle per train at one time is sufficient to cause a derailment. Hence, the number of potential buckles at a given segment can be reduced to the annual number of trains passing over the segment.

Alternatively, one can consider that only one buckling event is possible for a total track segment occupied by one train.

Defining

n_c = number of cars in a typical train
l_c = typical car length (expressed in miles)
n_t = number of trains in a typical day
n_p = number of potential buckling events possible
e_b = expected number of buckles per year

Then

n_p per mile = number of train lengths in 1 mile = $1/(n_c l_c)$
n_p per mile per day = $n_t/(n_c l_c)$
n_p per mile per annum = $365 \, n_t/(n_c l_c)$
e_b, expected number of buckles per year per mile = $P_a 365 n_t/(n_c l_c)$

Supposing e_b is required for a 100 mile segment, then

e_b, expected number of buckles per year for 100 miles = $P_a 365 \times 100 n_t/(n_c l_c)$
$e_b = 3.65 \times 10^4 \, P_a n_t/(n_c l_c)$
$e_b = (3.65 \times 10^4 \, P_b P_T n_t)/(n_c l_c)$ (Equation 5-2)

Equation 5.2 shows that the annual number of buckles for the 100-mile segment will increase with the following:

- The number of trains
- The probability of buckling at a given temperature (which depends on track parameters such as lateral resistance, rail neutral temperature or track geometry)
- The temperature frequency over T_c (i.e., more hot days will increase buckling)

The last term in the equation represents the length of the train. This indicates that shorter train increases the number of buckles versus a longer train. This is because only one buckle per train is considered as a potential event, and a 100 mile segment will have more train lengths for a shorter train than a longer one. As an illustration of the foregoing theory, the following data will be used.

n_c = 100 (number of cars per train)
n_t = 10 (number of trains per day)
l_c = 1/100 (car length in miles)

then equation 5-2 simplifies to

$$e_b = 0.365 \times 10^6 \; P_b P_T \hspace{4cm} (5\text{-}3)$$

Numerical Example

As an illustration, the annual rail temperature distribution is used as shown in Table 5-1. Using the table, the frequency distribution, P_T, can be constructed as a function of rail temperature, T_R. Figure 5-1 shows the resulting frequency distribution P_T for 2 °F intervals obtained by interpolation and normalization.

Table 5-1. Annual Rail Temperature Data

Air Temp (°F)	Rail Temp (°F)	Number of Days per Year
30	60	15
40	70	30
50	80	50
60	90	70
70	100	70
80	110	70
90	120	30
100	130	20
110	140	10

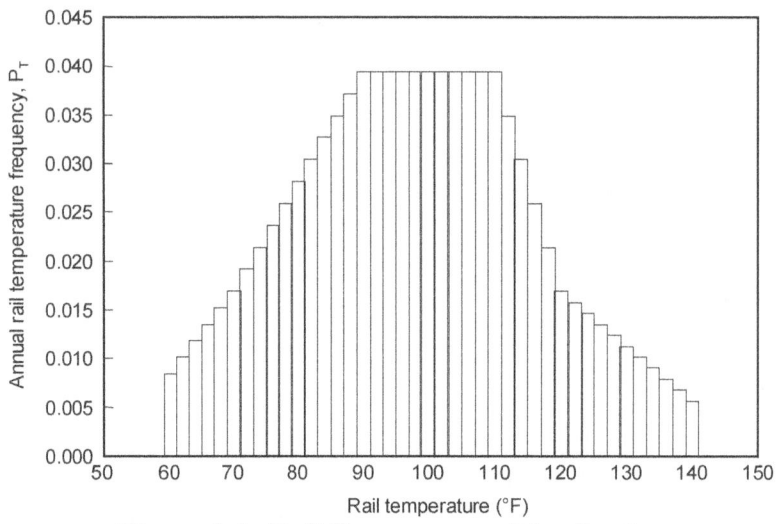

Figure 5-1. Rail Temperature Distribution

Baseline Case

As an example, consider a baseline case (Case A) for which the lateral resistance, neutral temperature, and the misalignment distributions are shown in Table 5-2. Figure 5-2 shows these distributions graphically. CWR-RISK takes the input misalignment and the lateral resistance data and formulates a distribution representing the strength of the track. Because the input data is an array with probabilities built in, the strength curve is also an array that is a function of probability. Inside the CWR-RISK program, this array is then interpolated to find values for every 1-degree increment. Next, the data is normalized so that all the probabilities sum to unity. This array represents the strength of the track represented by the allowable temperature, ΔT_{all}, which is obtained from the program.

Table 5-2. Input Distribution for Baseline Case A

Lateral resistance (lb/in)	Frequency (%)	Rail neutral temp (°F)	Frequency (%)	Misalignment amplitude (in.)	Frequency (%)
70	0.00	50	0.00	0.00	0.29
85	0.05	60	0.04	0.25	0.35
100	0.25	70	0.21	0.50	0.22
115	0.40	80	0.50	1.00	0.08
130	0.25	90	0.21	1.25	0.04
145	0.05	100	0.04	1.50	0.02
160	0.00	110	0.00	1.75	0.00

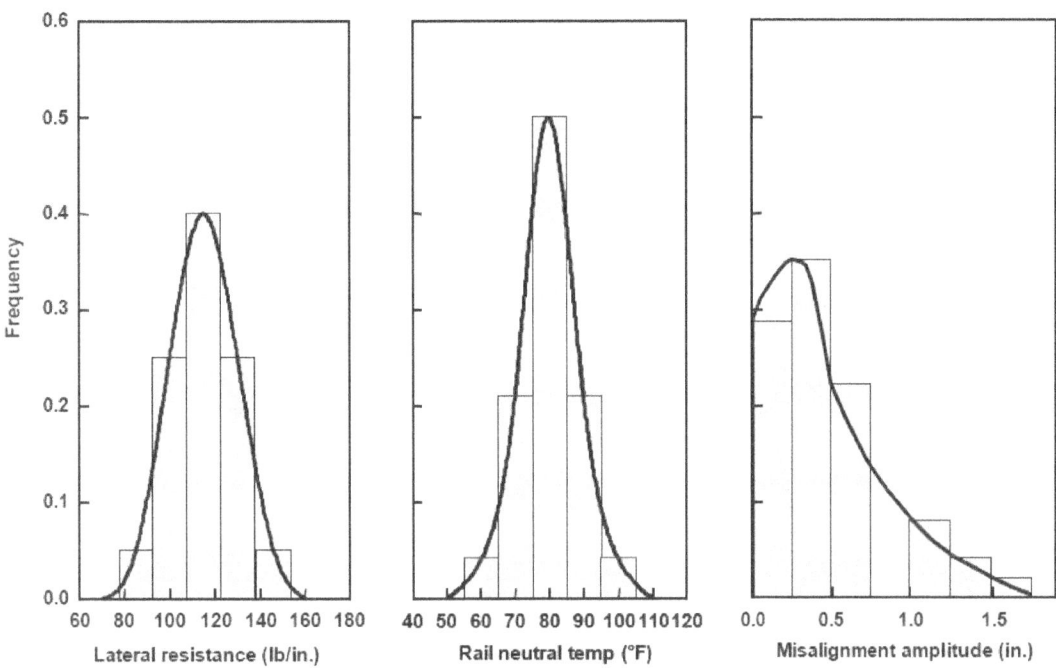

Figure 5-2. Distributions for Baseline Case A

As discussed in Chapter 3, CWR-RISK also calculates the load density curve. This is basically the rail neutral temperature minus the rail temperature (i.e., the rail temperature in degrees above the neutral temperature). CWR-RISK then overlays the load distribution curve onto the track strength curve and evaluates the buckling probability through the use of the convolution integral. The probability of buckling at a rail temperature is then output from the program. Figure 5-3 gives an example of a 5° curve with assumed track characteristics as in Table 5-3.

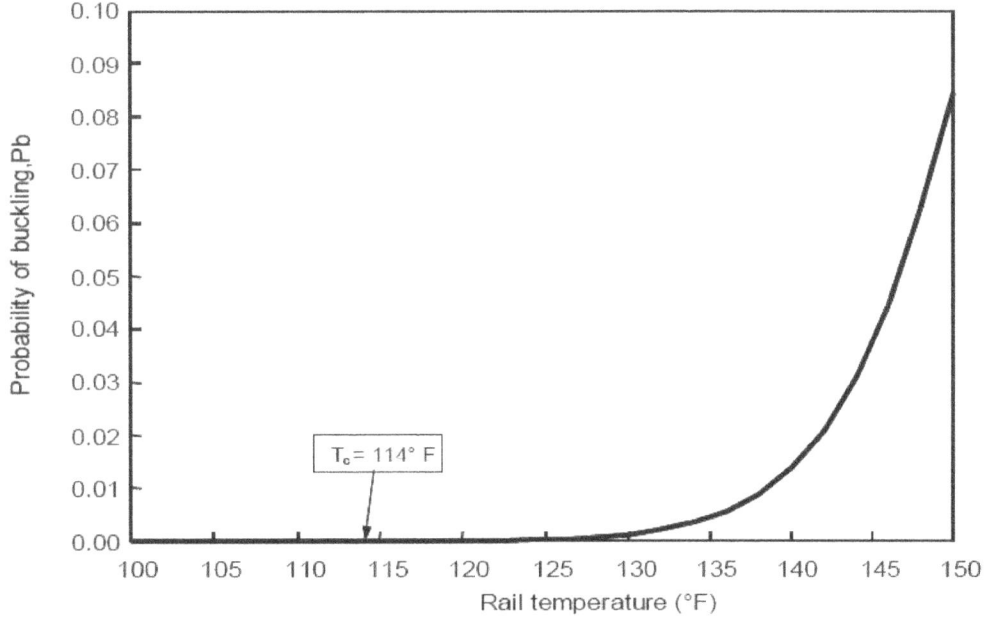

Figure 5-3. Buckling Probability at a Given Temperature, Case A

81

Table 5-3. Assumed Curved Track Characteristics

CWR-RISK Illustrative Example Inputs

Rail size (lb/yd):	AREA 136
Tie type:	Concrete
Tie weight (lb):	750
Tie spacing (in):	24
Track curvature (deg):	5
Ballast type:	Granite
Tie ballast friction coefficient:	0.86
Torsional resistance (in-kips/rad/in):	25
Longitudinal stiffness (psi):	200
Foundation modulus (psi):	10000
Vehicle type:	Hopper

Figure 5-3 presents the relationship between the buckling probability and the rail temperature. From this graph, one can identify the critical rail temperatures, T_c, at which the buckling probability just begins to be significant as defined in Section 3.1.3. This is the critical temperature (T_c) above which there is a more finite level of buckling risk. The critical temperature is 114 °F for Case A with the assumed parameters given in Table 5-3.

Figure 5-4, derived from Equation 5-3, gives the expected annual number of buckles over 100 miles of track as a function of rail temperature for the annual cycle in accordance with the P_T distribution of Figure 5-1. If vehicle operations are permitted for the full range of expected rail temperature up to 140 °F, the expected number of buckles per annum is 28. If the vehicle operations were to be stopped at a lower temperature, for example 130 °F, then the number of buckles will reduce to five as seen from Figure 5-4. Thus, the data in the figure can be used to identify the maximum permissible rail temperature for vehicle operations for a chosen maximum permissible annual number of buckles. This limiting temperature for train operation is defined as T_L. The dashed line shows a T_L = 130 °F if the permissible number of buckles per annum is five.

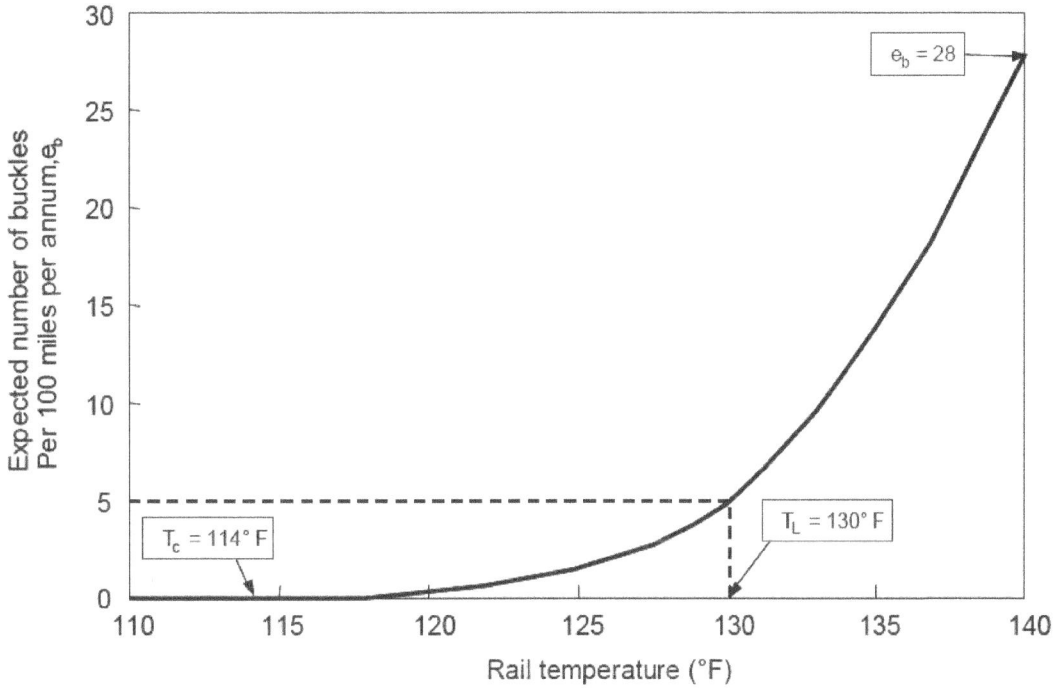

Figure 5-4. Number of Buckles versus Rail Temperature (Case A)

Slow Orders

As stated earlier, T_c represents the critical rail temperature below which the probability of buckling can be considered as zero; hence full normal line speed can be permitted for the vehicles up to this rail temperature. It has already been concluded that, for a chosen maximum number of permissible buckles in a year, a limiting temperature, T_L exists. At temperatures above this T_L vehicle speeds should be stopped (or slow ordered to 10 mph where practical). Given that the speed at T_c is the maximum permissible speed V_{max} and at T_L the speed is zero, the issue is to determine a permissible speed for the rail temperatures within this temperature regime of $T_c \leq T_R \leq T_L$. Since at present no consensus and common rationale for hot weather slow order policy determinations exists, a risk-based speed reduction formula is proposed here.

If the damage due to buckled track derailment can be assumed to be proportional to the vehicle kinetic energy and hence to the square of the speed, the suggested speed formula [13] for the same level of damage at all speeds is the following:

$$V_r/V_{max} = \{\ 1 - P_b(T)/P_b(T_L)\ \}^{0.5}$$

Equation 5-4

Where

  V_r = Reduced speed
  V_{max} = Permissible maximum authorized line speed
  $P_b(T)$ = Buckling probability at rail temperature, T
  $P_b(T_L)$ = Probability of buckling at the limiting temperature, T_L

For Case A, using the foregoing formula, the speed ratio is calculated and shown in Figure 5-5 for an assumed five or less annual buckles. As seen from the figure, the permissible speed falls off rapidly from its maximum value at T_c to zero at T_L. In practical applications, the graph may be idealized into a bilinear approximation by the lines GH and HI or any other linear combination in the gray zone.

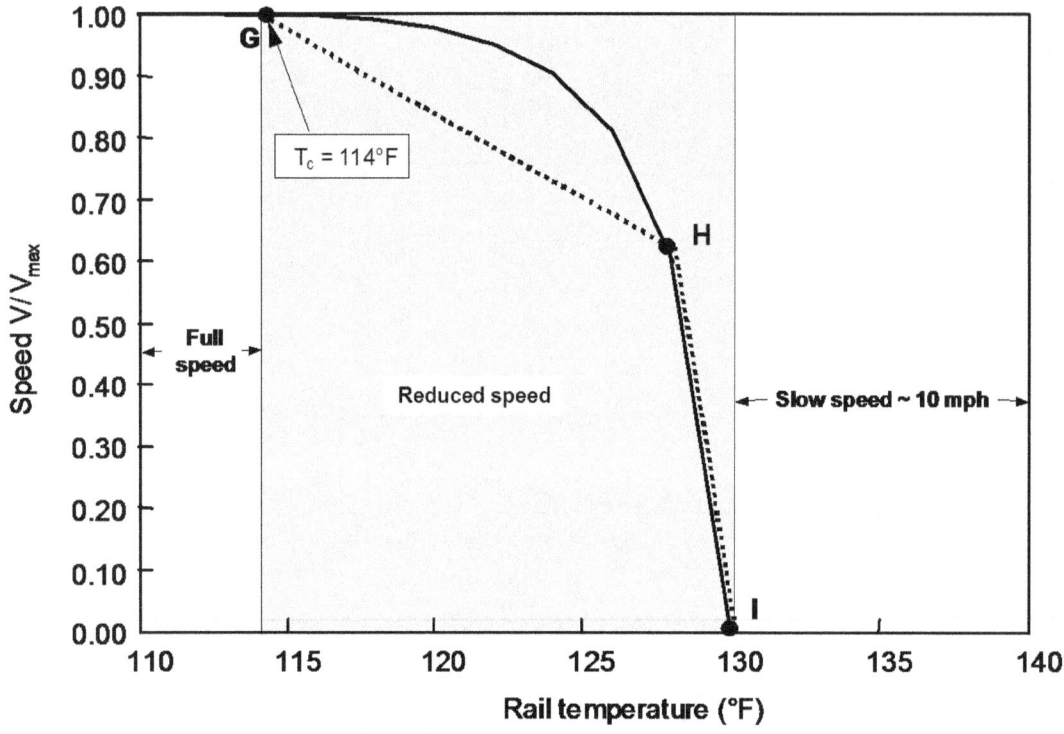

Figure 5-5. Permissible Speeds for Case A

Parametric Studies

5° Curve

To show the benefits of probabilistic analysis, this section presents a parametric study. As discussed in Section 5.1, the annual number of expected buckles depends mostly on how well the track is maintained in regard to the lateral resistance, the rail neutral temperature, and its alignment. The railroads can make choices in the maintenance of these three key parameters and attain the same level of buckling safety assurance (i.e., the same number of annual buckling events over a given territory). To illustrate this point, four possible cases of improvements in maintenance on the baseline track are considered, Case B, Lateral resistance, Case C, Neutral Temperature, Case D, misalignment amplitude, and case E improvement on all three parameters. Figure 5-6 through Figure 5-9 show schematically the potential improvements.

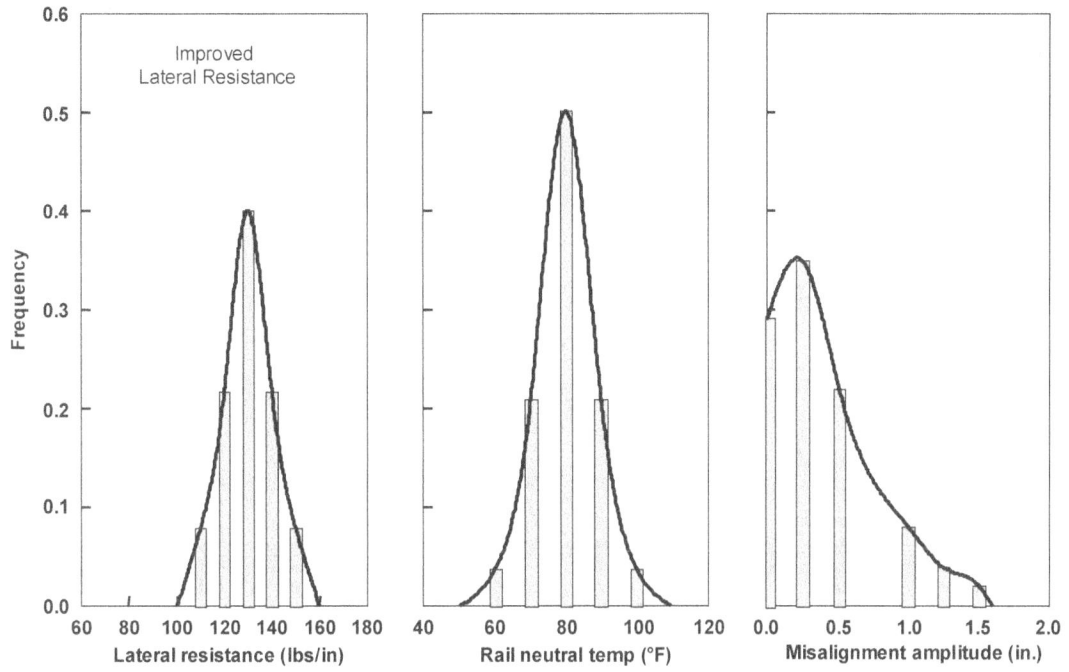

Figure 5-6. Distributions for Case B

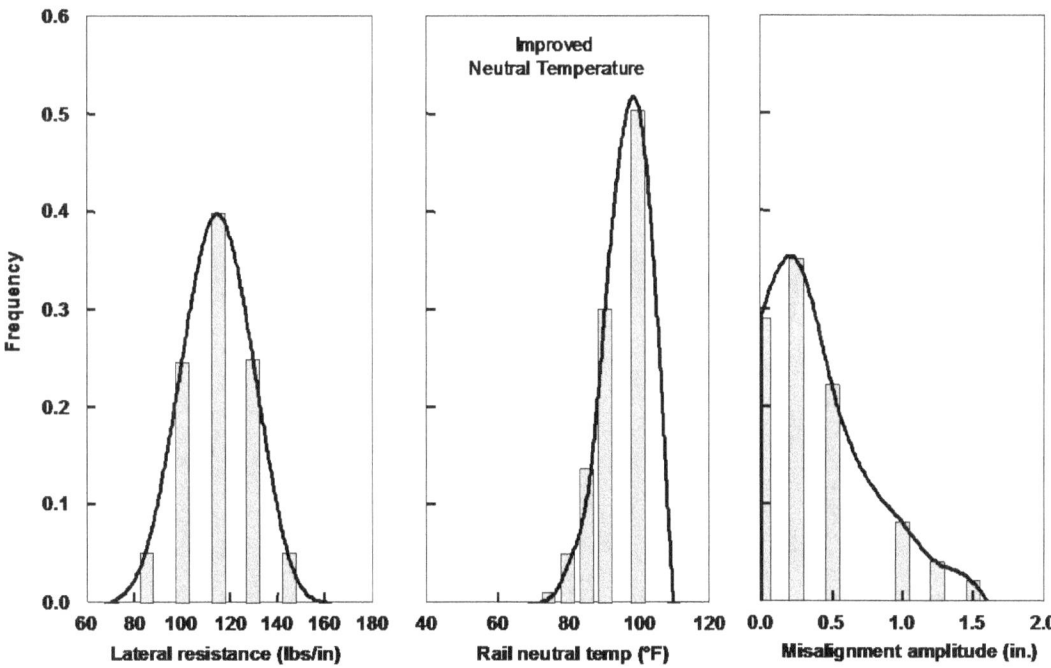

Figure 5-7. Distributions for Case C

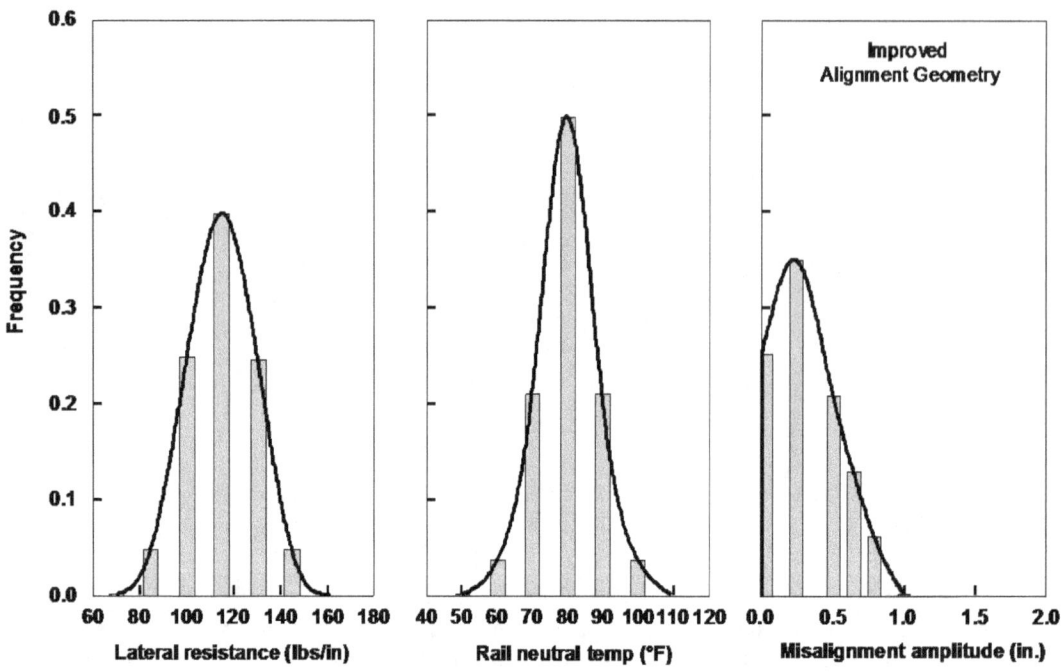

Figure 5-8. Distributions for Case D

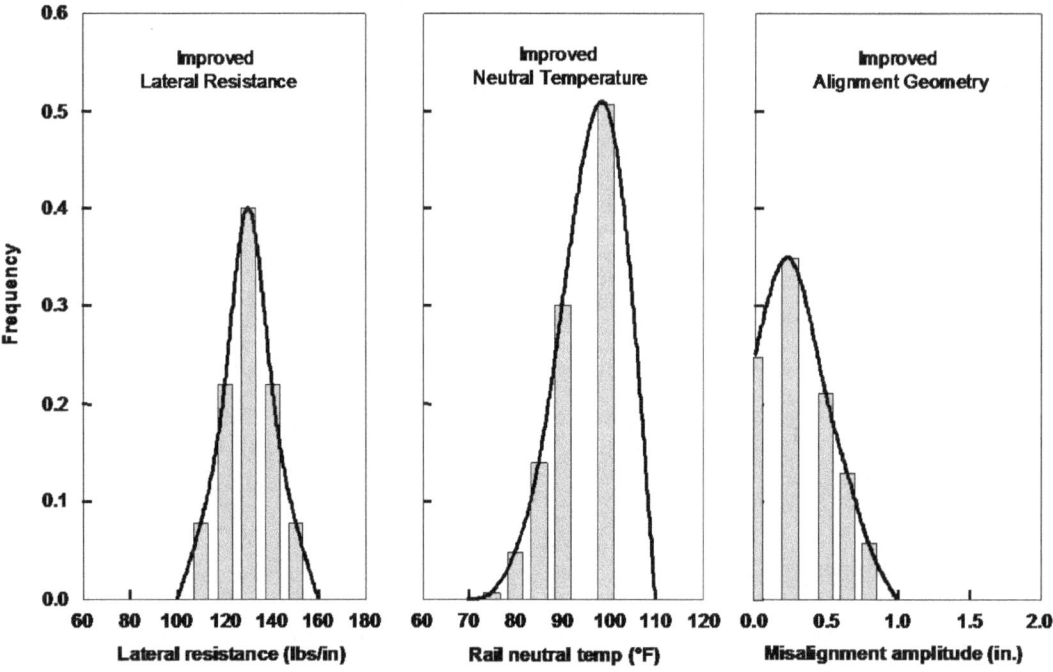

Figure 5-9. Distributions for Case E

Case B shows improvement in the lateral resistance distribution by having full cribs, maintaining a good shoulder geometry, and ensuring high levels of ballast consolidation. The result of such measures will be the better distribution of lateral resistance with an increase in the mean value and the range as compared to the resistance distribution shown in Figure 5-2 (baseline Case A). The other two distributions of the neutral temperature and misalignment are the same as in Case A.

Case C has an improved neutral temperature distribution when compared to the baseline Case A. This can be achieved by proper restressing of rails after rail repair and after track realignment, maintaining effective rail anchors and high toe loads on elastic fasteners, and minimizing curve breathing in wintertime, using adequate ballast inside and outside the shoulder and monitoring the curve position. The distributions of lateral resistance and misalignment in Case C are the same as in the baseline.

Case D represents improvements in the alignment over the baseline. This improvement can be accomplished through more frequent geometry inspection and alignment corrections. The generation and growth of misalignments can also be mitigated by reducing vehicle lateral forces and high cant deficiency curves. The distributions of the lateral resistance and the neutral temperature in Case D are the same as in the baseline case.

Finally, Case E, where all three improvements occurring simultaneously are incorporated; this can represent an ideal (strong) track as shown in Figure 5-9.

Clearly, one would expect the baseline to result in the highest number of buckles per year and the ideal case to result in the lowest number of buckles. The question of interest is how well the relative improvements will perform with respect to buckling safety and their relative cost of implementation. The performance benefits will be quantified using the probabilistic theory, but the cost benefit analysis will not be presented because it depends on individual railroad practices and procedures.

Performance Comparisons

Figure 5-10 through Figure 5-13 show the expected number of buckles and the annual probability of buckling as a function of rail temperature for the first four cases. Table 5-4 summarizes salient results for all cases. The results include the critical temperature T_c, the limiting temperatures for the number of buckles under 5 per annum, and the expected number of buckles if normal vehicle operations speeds are continued up to $T_R = 140°$ F. The dramatic decrease in the number of buckles at 130° F, for example, compared to 140° F operations can be clearly seen. An inspection of the results for all the cases at 140° F shows that the neutral temperature distribution improvements will also bring significant reduction in the annual number of buckles. At 130° F, the neutral temperature distribution improvement alone may be adequate to eliminate all the potential buckles under the assumed conditions. The improvements in all three parameters (Case E) will result in zero buckles even at 152° F. A figure for Case E is not shown because of the zero number of expected buckles up to 152° F.

87

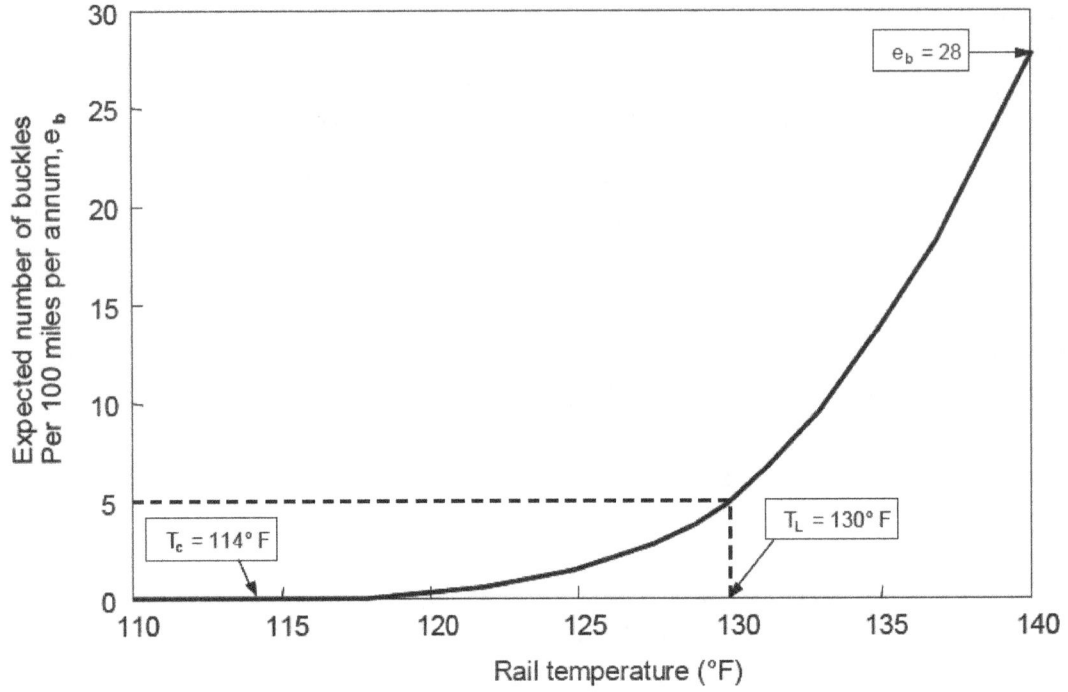

Figure 5-10. Number of Buckles versus Rail Temperature (Case A)

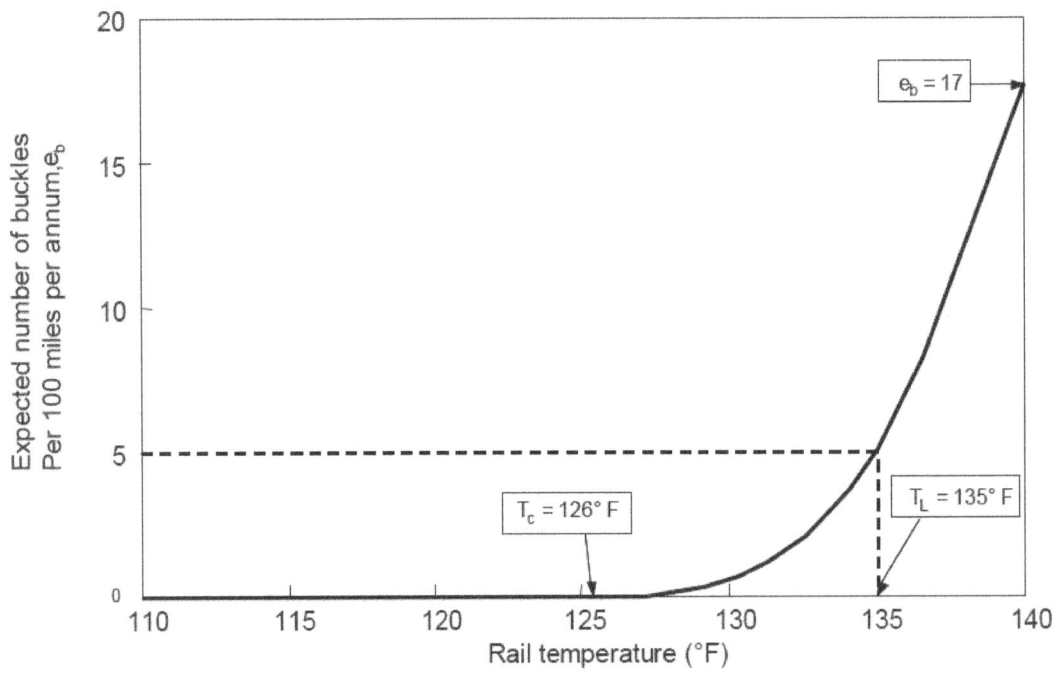

Figure 5-11. Number of Buckles versus Rail Temperature (Case B)

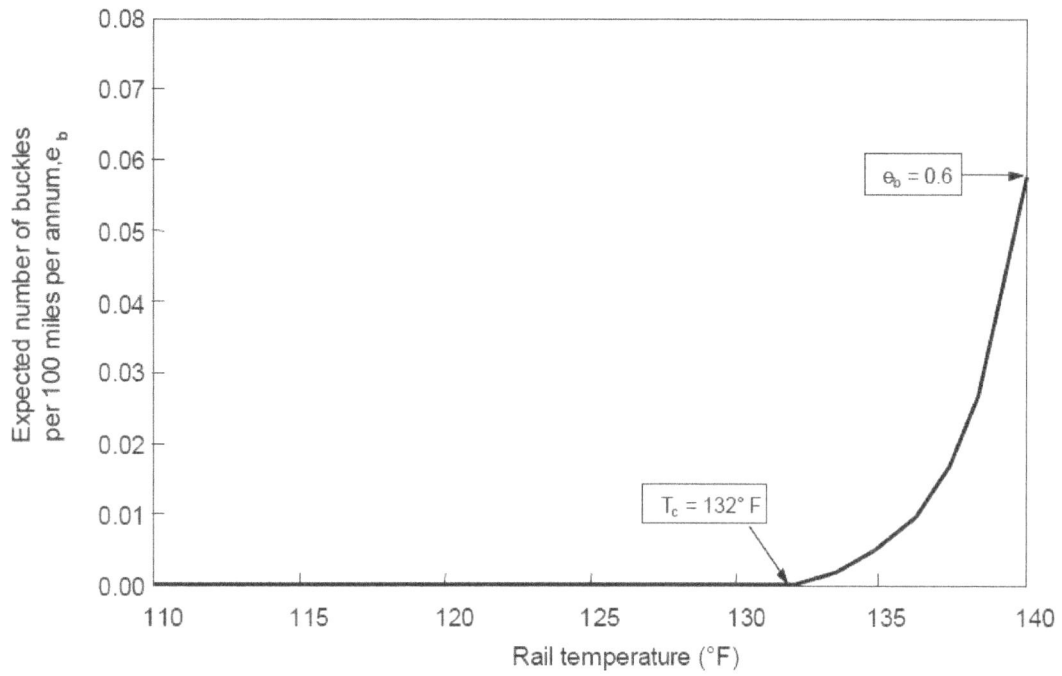

Figure 5-12. Number of Buckles versus Rail Temperature (Case C)

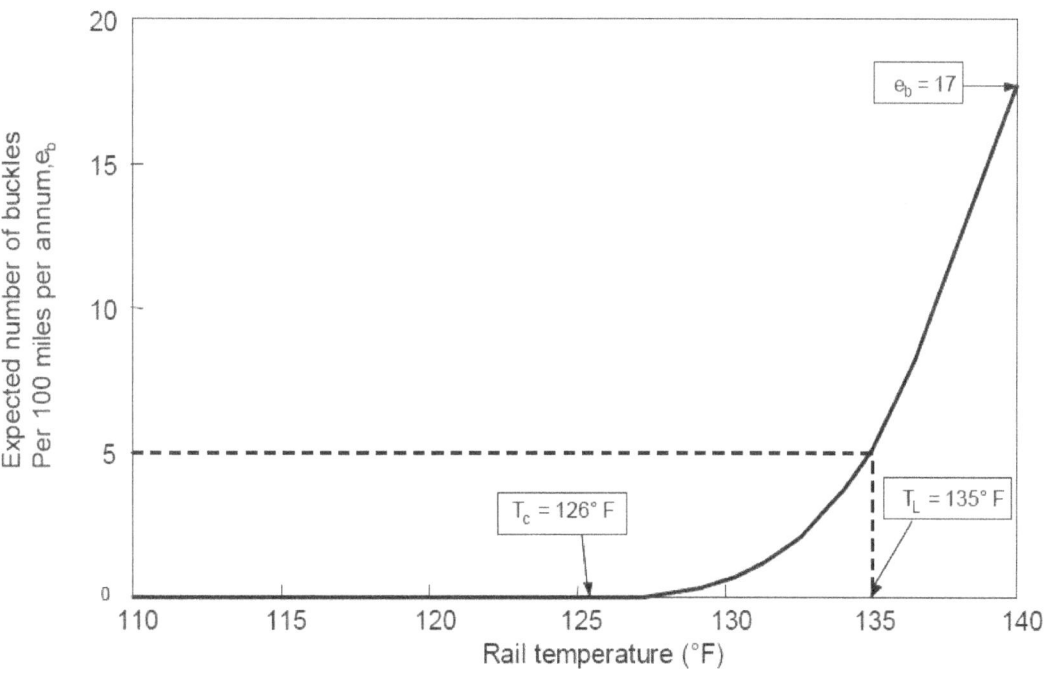

Figure 5-13. Number of Buckles versus Rail Temperature (Case D)

Table 5-4. Comparison of Track Cases

Case	Description	T_c	T_L for 5 buckles per annum	Number of buckles per annum at $T_R = 140\ °F$	Number of buckles per annum at $T_R = 130\ °F$
A	Weak Baseline Track	114 °F	130 °F	28	5
B	Improved Lateral Resistance	126 °F	136 °F	12	0.66
C	Improved Rail Neutral Temperature	134 °F	>140 °F	0.06	0
D	Improved Alignment	126 °F	135 °F	17	0.56
E	Improvements In All Three Parameters	152 °F	>152 °F	0	0

Tangent Track

Analysis has also been conducted in which the baseline case parameters (Case A) are applied to tangent track to study how track curvature affects the number of buckles per year. Figure 5-14 shows the results for tangent track.

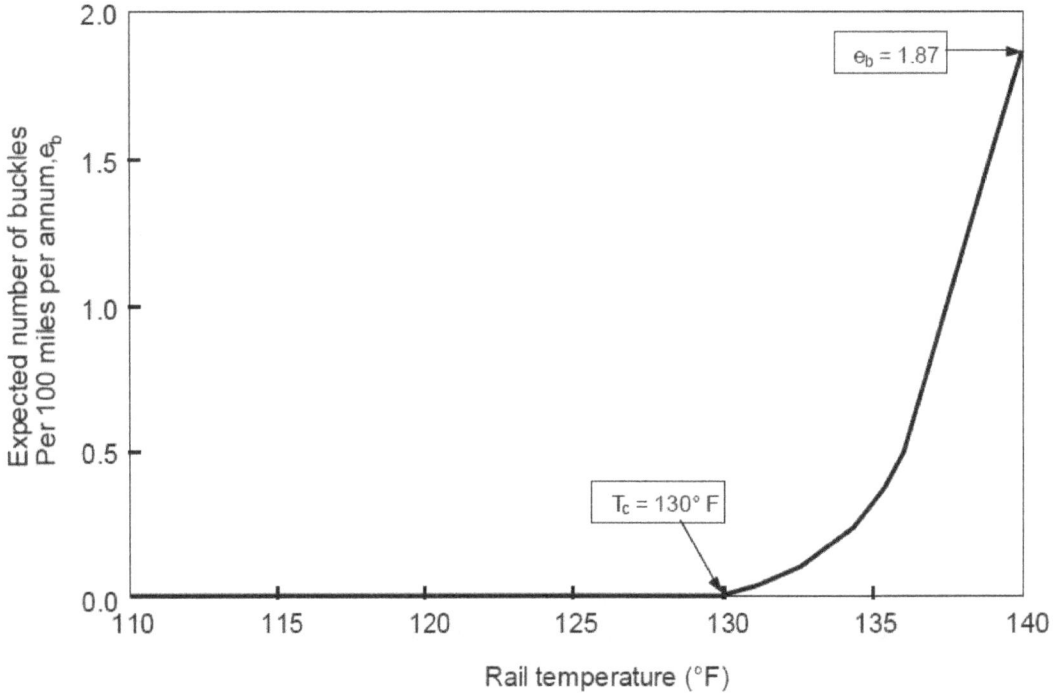

Figure 5-14. Number of Buckles versus Rail Temperature for Tangent Track (Case A)

Applying the Case A parameters to tangent track as opposed to 5° track increases T_c from 114 °F to 130 °F. In addition, the number of annual buckles decreases from 28 to less than 2 for every 100 miles per annum, showing that curved tracks are more vulnerable to buckles as compared to tangent tracks.

Slow Orders Parametric Study

Figure 5-15 shows the permissible speed ratios (V_r/V_{max}) with respect to rail temperature for each of the cases of the parametric study. These graphs are derived using the equation in Section 5.1.4. As expected, the baseline Case A indicates quick reduction in speed at lowest temperatures. With improvements, the curves show higher speeds at relatively high temperatures. Case C performs far superior than case B and case D and allowing a vehicle to maintain full speed up to 140 °F. This means that an effective way of maintaining high-speed operation at high rail temperatures is through the assurance of a better (higher) rail neutral temperature condition. The ideal Case E outperforms the individual cases and allows full speed up to 152 °F.

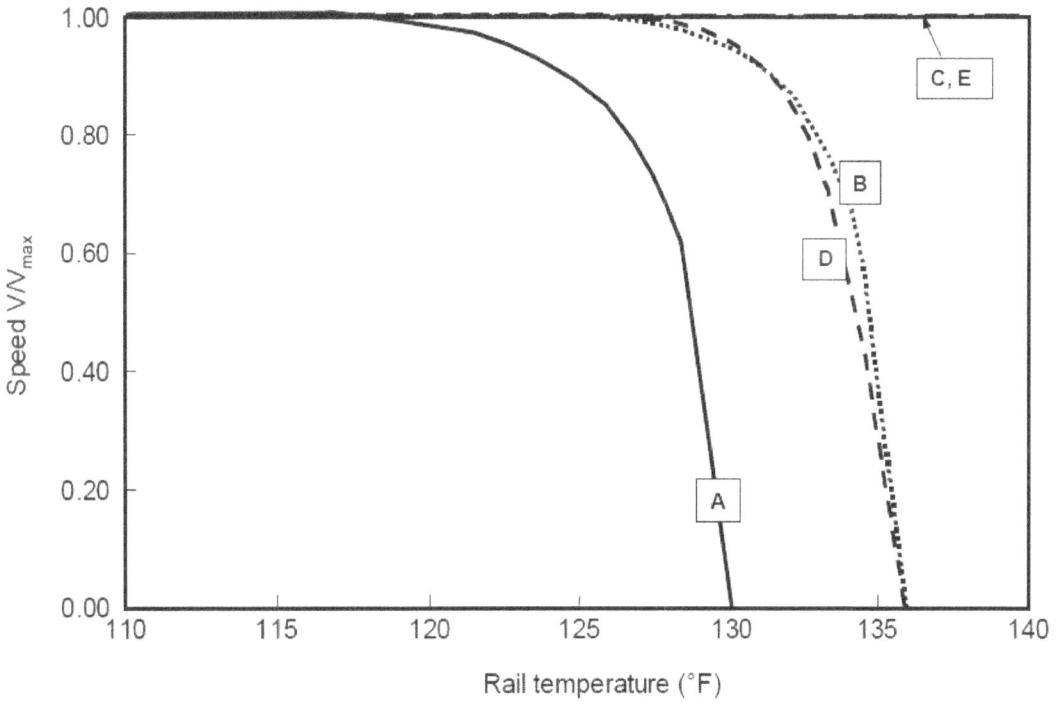

Figure 5-15. Permissible Speeds for All Cases

Summary

- By determining the distributions of the key input parameters through measurements, railroads can optimize their maintenance strategies for minimization of buckling accidents and associated costs.
- The probabilistic buckling theory provided in CWR-RISK predicts the buckling probability of CWR track at any given rail temperature. The graph relating the buckling

probability and the rail temperature depends largely on the three key parameters of lateral resistance, neutral temperature, and misalignment. These three parameters are expressed in terms of statistical distributions. The buckling probability graph identifies the critical temperature at which buckling probability becomes significant and can define slow order requirements at elevated temperatures.

- From the probability versus rail temperature curve, the expected number of buckles per annum over a given track length can be predicted, if the annual temperature distribution of the rail is known. This information permits more practical measures for buckling safety assurance through targeted track maintenance operations and slow orders.

- A limiting rail temperature for vehicle operations can be identified from the analysis, based on the accepted number of buckles per annum over a given territory.

- The influence of alternate maintenance strategies on buckling safety, such as control of ballast resistance, neutral temperature, and track alignment or limiting the vehicle operations at lower temperatures can be evaluated. This will allow for performing cost-benefit analyses on maintenance procedures to reduce the buckling related damages.

- A risk-based theory is proposed for slow order management. The theory predicts the critical temperature to initiate slow orders and the appropriate speed reduction up to a limiting temperature.

6. Safety Aspects of CWR

This chapter presents CWR buckling safety aspects, including safety limits, test validation of the limits, and implementation methodologies of the safety limits for buckling prevention and mitigation.

Safety Limits

As discussed in the previous chapters, two approaches to the safety evaluation of CWR track exist. One is the deterministic method, which focuses on the safe allowable temperature for a given set of track conditions and vehicle parameters. When the rail temperature exceeds this allowable temperature, vehicle operations should be curtailed. The second method is the probabilistic method, which gives a finite probability of buckling at any rail temperature exceeding the critical temperature, T_c. This method can also calculate the expected number of buckles per annum over a given track section length and requires annual rail temperature distribution information. When the rail temperature exceeds T_c, vehicle operations can be continued at reduced speeds as discussed in Chapter 5.

The two methods should not be considered as alternative choices in the management of CWR safety. Rather, they complement one another in overall safety assurance and provide reasonable flexibility for railroads use and implementation. The deterministic method is appropriate for specific track sections with a history of buckling incidents and for day-to-day buckling safety assurance on chosen sections of track. The probabilistic method provides a global approach for buckling safety improvement for extended periods of operations over a larger CWR territory.

Deterministic Method

The allowable rail temperature is the sum of the safe allowable temperature increase and the CWR neutral temperature. The safe limits can be presented in one of the following schemes:

1. Application of CWR-BUCKLE
2. Use of graphics/charts prepared for a range of track conditions/parameters via CWR-BUCKLE
3. Uncoupled approach

The following paragraphs discuss these in detail.

1. Application of CWR-BUCKLE

The safety criteria and evaluation methodology illustrated in Chapter 2 can be applied directly by exercising CWR-BUCKLE. With the pertinent track and vehicle input parameters, the program calculates the allowable temperature increase and performs the safety analysis. The required controlling parameters for buckling safety, such as lateral resistance and rail neutral temperature, are used to develop the safety criteria.

The advantage of this method is its ability to account for all the parameters in the system and provide a quick calculation of allowable temperatures. The disadvantage is that it requires some knowledge of the required input parameters and some expertise in running the program.

2. Use of Graphs/Charts

In this method, charts are prepared giving the maximum permissible temperature for buckling safety in terms of two the two primary parameters: lateral resistance and CWR neutral temperature. Lateral resistance is chosen as a primary variable in the determination of the safe allowable temperature increase with the other (secondary) parameters set at their average or nominal values. For an assumed line defect and track curvature, the maximum allowable rail temperature can be plotted against the lateral resistance over a range of neutral temperature.

Figure 6-1 through Figure 6-4 constitute examples of safety limit charts for 136# rail CWR concrete tie track with Tangent, 3°, 5°, and 7° curvature for an FRA Class 6 line defect of 0.5 in, while Figure 6-5 through Figure 6-8 are examples for Class 4 line defect of 1.5 in. The assumed fixed parameters for the four track conditions and the two classes of track are as shown in the inset in Figure 6-1.

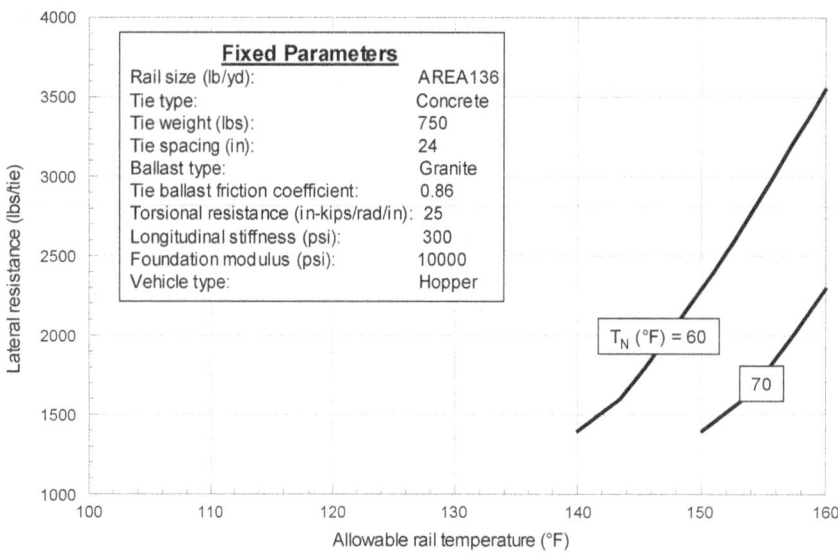

Figure 6-1. Buckling Safety Limits Chart for Tangent with 0.5" Line Defect

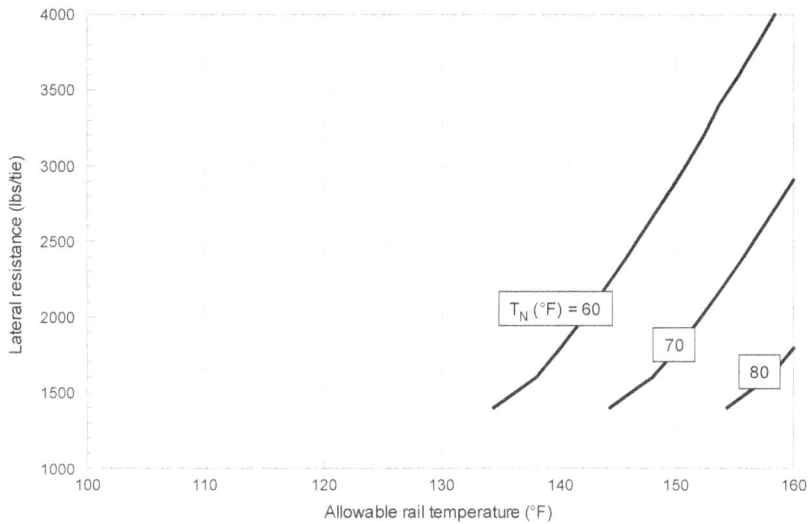

Figure 6-2. Buckling Safety Limits Chart for 3° Curve with 0.5" Line Defect

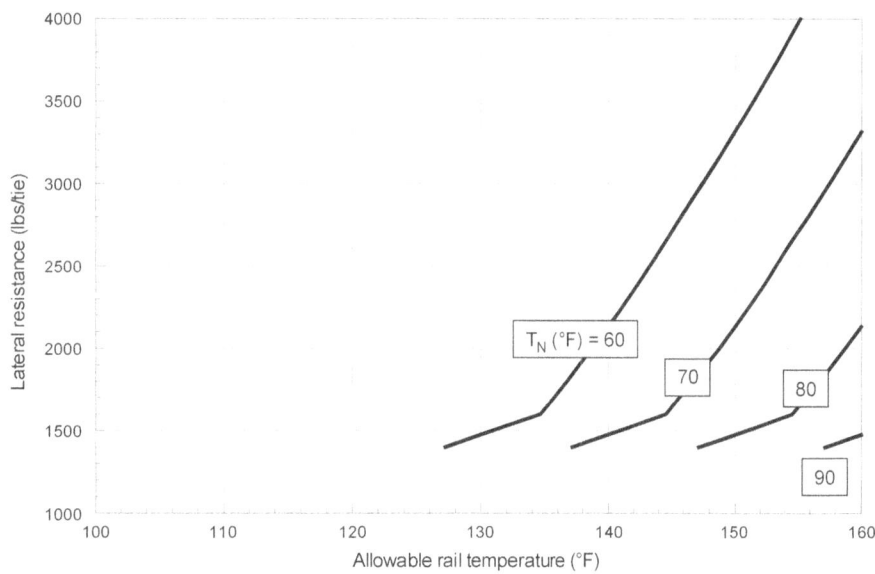

Figure 6-3. Buckling Safety Limits Chart for 5° Curve with 0.5" Line Defect

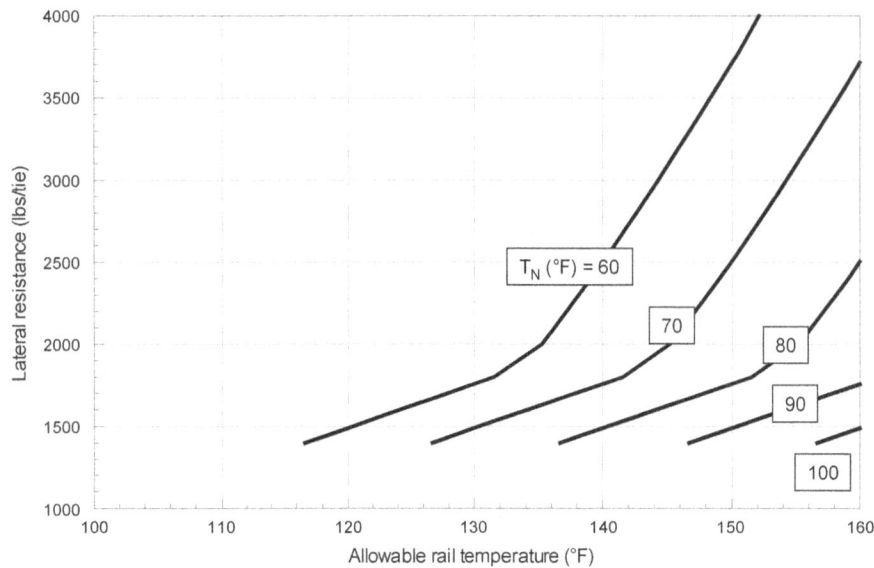

Figure 6-4. Buckling Safety Limits Chart for 7° Curve with 0.5" Line Defect

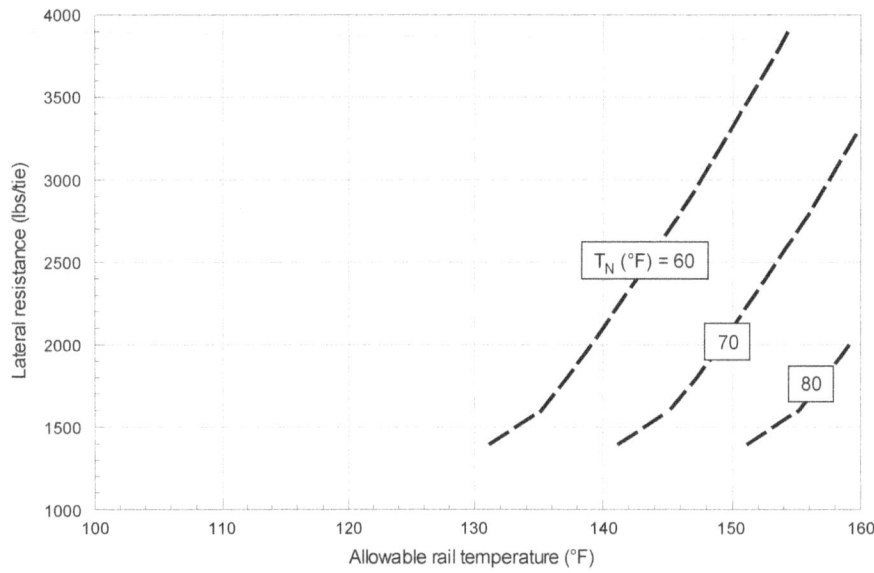

Figure 6-5. Buckling Safety Limits Chart for Tangent with 1.5" Line Defect

96

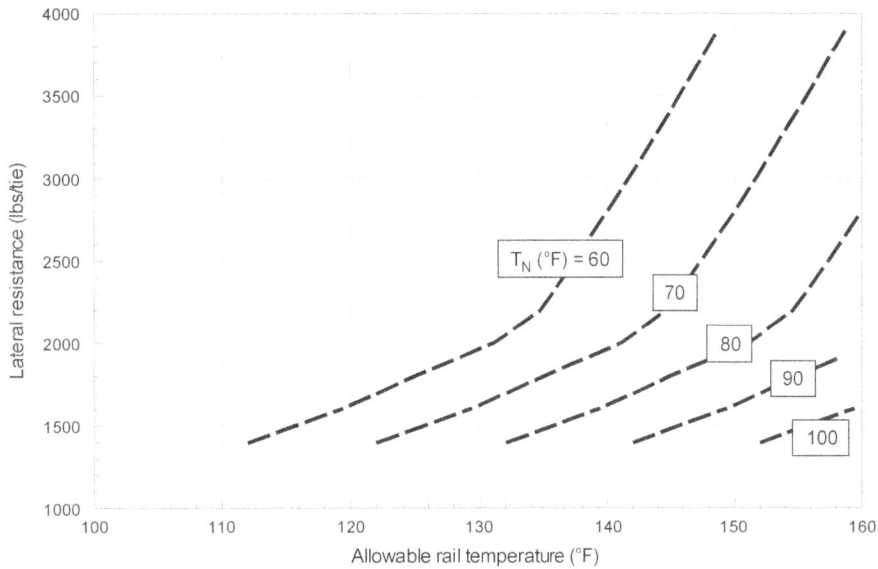

Figure 6-6. Buckling Safety Limits Chart for 3° Curve with 1.5" Line Defect

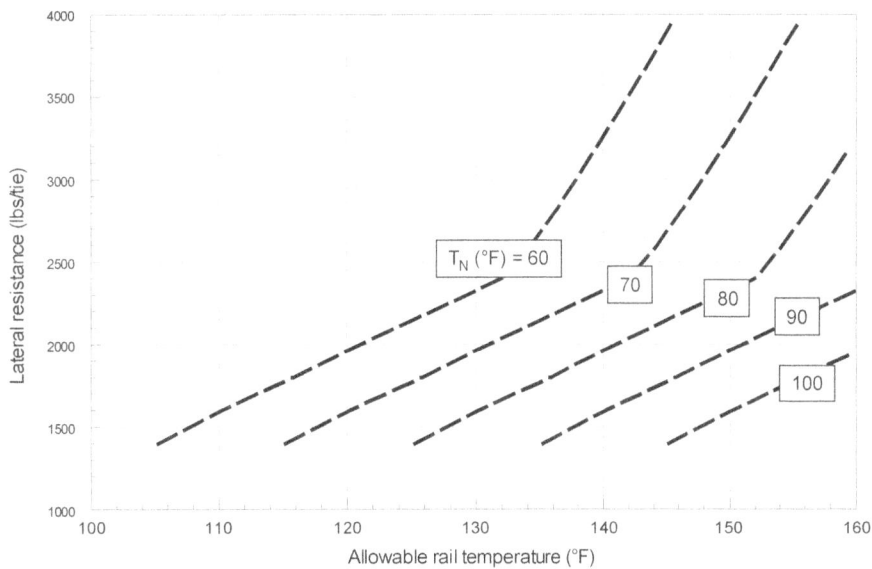

Figure 6-7. Buckling Safety Limits Chart for 5° Curve with 1.5" Line Defect

97

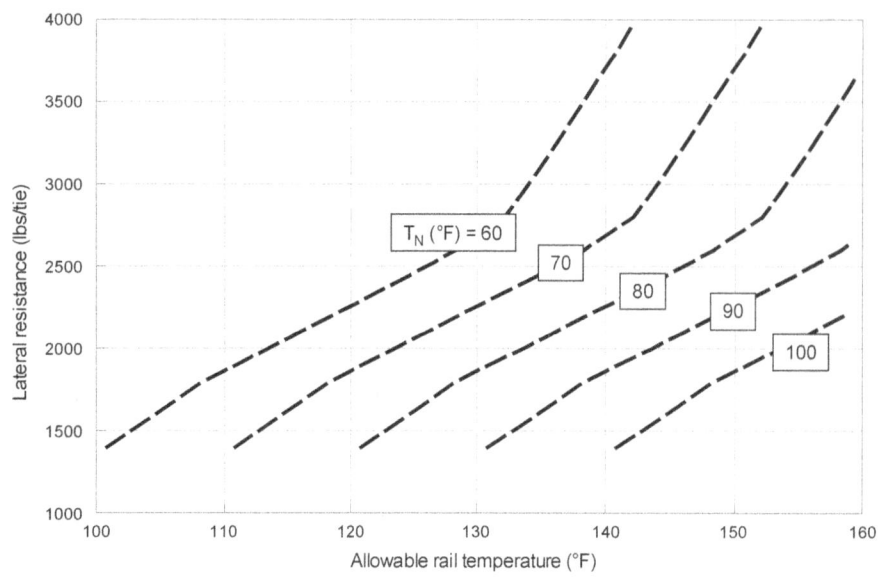

Figure 6-8. Buckling Safety Limits Chart for 7° Curve with 1.5" Line Defect

The advantage of these graphical representations is that the safety limits can be easily read off the graphs for implementation without the need to run CWR-BUCKLE. The disadvantage is that graphs must be fairly extensive to cover all track parameters in addition to lateral resistance, (e.g., different rail size, wood tie tracks, and fastener resistances). The parametric study result of Chapter 4 can also be invoked to help in gaining a better insight to parametric influences of the safety limits.

3. Uncoupled Approach

The safety limits in the charts for the maximum allowable rail temperature are derived from the combined effects of the allowable rail temperature increase (which is dependent on lateral resistance) and the neutral temperature. For a given maximum rail temperature at a given geographical region, one can determine the minimum required lateral resistance for a possible lowest neutral temperature. For example, referring to the previous figures, the required lateral resistance can be calculated on the assumption that the neutral temperature drops off to a value between 60 °F and 70 °F from its intended value of 95 °F or higher. From the safety limit charts, Table 6-1 shows the required values of lateral resistance in a region with 140 °F maximum rail temperature.

Table 6-1. Minimum Resistance Values for Neutral Temperature of 60 °F and Rail Temperature of 140 °F

Curvature (Degrees)	Required Lateral Resistance (lb/tie)	
	0.5" Line Defect	1.5" Line Defect
Tangent	1400	2100
3	1800	2800
5	2100	3200
7	2500	> 3500

98

In this table, the neutral temperature is independently set at a minimum limit of 60 °F. For curvatures of 7° and greater, the minimum resistance values required will be over 3500 lb/tie. Such high values are generally difficult to ensure consistently because of maintenance activities, and curve movement. However, if a limit of 70 °F is stipulated for the neutral temperature, the resistance values fall into the range as shown in Table 6-2.

Table 6-2. Minimum Resistance Requirement for Neutral Temperature of 70 °F and Rail Temperature of 140 °F

Curvature (Degrees)	Required Resistance (lb/tie)	
	0.5" Line Defect	1.5" Line Defect
Tangent	<1400[1]	<1400[1]
3	<1400[1]	1950
5	1500	2300
7	1800	2700
[1] Although resistances below 1400 lb appear in Table 6-2, such low values are not recommended in practice since they may give rise to other problems such as track shift/curve breathing and general instability under high vehicle loads.		

These tables show that lateral resistance and neutral temperature can be independently specified to obtain the allowable rail temperature values.

Probabilistic Approach

This approach is useful in planning overall CWR track maintenance to control the buckling incident rate within a certain limit, such as less than one buckle per every 500 miles per annum on a given territory. Because the thousands of miles of railroad track exhibit variable conditions, the probabilistic approach is more appropriate for overall buckling safety evaluations. The probabilistic approach predicts a critical temperature, T_c, above which buckling risk exists. The method can also predict the number of buckles in the territory that can occur annually for a rail temperature in the range of T_c and T_{MAX}. The maximum permissible temperature for vehicle operations, T_L, which is equal or less than T_{MAX}, can be determined for a specified limit allowed on the permissible number of buckles.

This method will also provide a speed reduction formula for vehicle operations when the rail temperature exceeds the critical temperature T_c. The speed reduction formula provides a rational basis for hot weather slow order practices.

In summary, the method specifies the following:

T_c = critical temperature above which buckling risk exists
N_B = maximum number of permissible buckles per annum per section of track
T_L = maximum permissible rail temperature to limit the number of buckles below N_B

As stated in Chapter 5, the value of T_c can be derived using the frequency distributions of the lateral resistance, neutral temperature, and misalignment amplitude. The limit values in these distributions, (i.e., the lowest values of lateral resistance and neutral temperatures together with the highest value of the misalignment amplitude) can also be used in the deterministic approach to evaluate T_c. This is because the CWR-RISK predicted T_c value is the same as the T_{all} in the deterministic approach of CWR-BUCKLE for those specific input parameters.

The maximum permissible number of buckles, N_B, is to be specified based on the railroad's risk acceptance in terms of potential derailment impact, possibly in conjunction with historical data on buckled track derailments for the particular territory. Table 6-3 shows one hypothetical specification and is intended for illustrative purpose only. The T_c numbers can be determined from CWR-RISK application to specific territory parameters/conditions, while the maximum permissible buckles from historical records and railroad risk acceptance.

Table 6-3. Example Specifications on a Territory for Normal Freight (no Hazmat)

Track Class	Minimum T_c °F	Permissible Number of Buckles per 100 Miles per Year
3	120	.1
4	130	.05
5	135	.01
6	140	.001

In addition, Table 6-3 shows the specifications on the critical temperature T_c, which represents the rail temperature at and below which the buckling probability is zero (10^{-6}). To preserve this T_c, the lowest values of the lateral resistance and neutral temperature (measured or stipulated for the territory and used in the analysis to determine T_c) must be maintained. T_c should increase for higher track classes (assuming similar values for lateral resistance and neutral temperature), for which the annual buckling rates should decrease. Depending on the assumptions of the maximum permissible buckles, operation can be performed above the stipulated T_c (i.e., to a level T_L as inferred in Chapter 5).

Safety Limit Validations

Deterministic Limits

The buckling theory was validated with full-scale tests as discussed in Chapter 3. The limits as derived in the Deterministic Method were also validated by direct full-scale tests [6, 7] on the CWR tracks at TTC in Pueblo, CO.

In the tests designed to validate the safety limits, two tangent track tests, two 5° curved track tests, and one test on a 7.5° curve were conducted. In each case, the rails were heated artificially, and multi cars consist made several passes at rail temperatures near and above the safety limit values. The tangent and 5° curved tracks showed only a very small growth in misalignments and were considered essentially stable. The 7.5° curve withstood the longitudinal forces generated at the safe allowable temperature but buckled out explosively at about 10 °F above the safe allowable temperature. This is consistent with the minimum margin of safety of 10 °F built into the prescribed safe allowable temperature.

Referring to Table 6-4, tangent tracks were tested up to temperatures slightly higher than the safe allowable. The highest temperatures for the 5° curved track tests were limited because of prevailing cloudy and windy conditions on the specific days of testing.

Table 6-4. Buckling Safety Limit Test Results

Test #	Curve	Peak Tie Lateral Resistance (lb/in.)	Misalignment		ΔT °F		Max. Vehicle Speed (mph)	Number Cars	Final Misalignment Amplitude (in.)	Comments
			Amplitude (in.)	Length (in.)	Theory	Test				
1. Tan I	0	68	0.88	335	84	92	20	48	0.98	No Buckling
2. Tan II	0	79	0.81	319	86	100	55	67	0.82	No Buckling
3. Curve I	5°	83	0.55	307	77	70	40	63	0.55	No Buckling
4. Curve II	5°	98	0.7	311	80	79	40	52	0.84	No Buckling
5. Curve III	7.5°	89	0.75	315	52	62	34	24	3.94	Track Buckled Under Train

Safety Limit Implementation

This section will describe the methods used to implement the deterministic limits. A formal methodology for the implementation of the probabilistic approach and its extension to risk-based safety performance standards is currently under development.

As stated previously in Section 6.1, the implementation is facilitated if the uncoupled approach is followed. For a given geographical region with known maximum rail temperature at and above which vehicle operations will be stopped (or are performed at a very slow speed), a set of required limiting values can be established:

- Lateral resistance $F_P \geq F_P{}^1$
- Neutral temperature $N_t \geq N_t{}^1$
- Misalignment amplitude $\delta_0 \leq \delta_0{}^1$
   [1] Limiting value for each parameter given in Table 6-1 and 6-2

Among the three limiting parameters, the neutral temperature is the most difficult one to monitor and control. It is expedient, therefore, to assume a minimum value for it and attempt to control other parameters at the required values. Although the neutral temperature is usually set during CWR installation between 90 °F to 115 °F (or even higher depending on the geographic location of the CWR), it can drop to much lower values for the several reasons discussed in previous chapters. It becomes appropriate to assume as a safety factor a reduced value for neutral temperature such as 60 °F or 70 °F. On this basis, the corresponding minimum required value for the lateral resistance can be determined from the charts in Figures 6-1 to 6-8 for line conditions as given in the insert of Figure 6-1. Indeed, Table 6-1 and Table 6-2 values for the minimum required lateral resistance were deduced for assumed minimum neutral temperatures of 60 °F and 70 °F, respectively, and serve as an example for the determination of the $F_P{}^*$, $N_t{}^*$, and $\delta_0{}^*$ values.

The CWR safety implementation is to be carried out through monitoring and controlling of the foregoing parameters within the limits indicated by the starred quantities. The following paragraphs discuss some procedures/methods for monitoring and controlling these parameters.

Lateral Resistance

Monitoring Hardware/Software

Monitoring lateral resistance at limited critical locations can be carried out using the STPT method described in Appendix B in accordance with the measurement requirements in Appendix D. In some cases, CWR-INDY's track lateral resistance predictor (see Equation 3.5 in Chapter 3) can be used to estimate the lateral resistance. The current STPT fixture can be drastically simplified in size and operation by requiring capture of only the peak resistance values which occur within small (0.3") displacements. The current method's large displacement measurement, the X-Y plotter, and the electric pump are not required; a simple hand pump is adequate to operate the hydraulics to produce the required loads to move the tie.

Monitoring Locations

Some of the critical conditions at a given location include the following:

- Recently maintained sections (surfaced, realigned, tamped, or after other operations, such as ballast and tie renewal)
- Substandard ballast section identified visually to have inadequate ballast
- Sections which have undergone dynamic track stabilizer (DTS) and/or traffic consolidation

Monitoring Methodology

As explained in Appendix D, the resistance is identified as the average for a 50 foot cell, which is chosen because it is representative of a buckle length under a car. One measurement per 50 foot concrete tie track and two measurements for wood tie track are deemed adequate to evaluate the resistance within a prescribed tolerable error as discussed in Appendix D.

The lateral resistance data will be examined and compared with the required values as indicated by the safety limits in the uncoupled approach. Those cells with resistance values less than the stipulated values in the safety limits must be restored to the safety limits and above using one of the following methods.

Measures to Modify and Control Lateral Resistance

Track lateral resistance can be controlled at and above a specified value by one or more of the following methods, including, full ballast-section, traffic consolidation, and mechanical consolidation.

Full Ballast Section: Good quality ballast, such as granite with good particle size and interlocking capability, full cribs and shoulder widths of at least 18 in would contribute to providing increased lateral resistance. Heavier concrete ties with roughened bottoms and scalloped designs are also useful in increasing the lateral resistance.

Traffic Consolidation: The ballast must be consolidated after any maintenance activity such as surfacing, tamping, and realignment. An empirical relationship exists between the lateral resistance and the traffic tonnage (MGT) as given by Equation 3-5 in Chapter 3. However, consolidation by traffic usually needs some time to accumulate to the level (usually around 0.1 MGT) required due to the typical low speed application. To ameliorate this, other means of consolidation is performed as discussed below.

Mechanical Consolidation: The ballast can be consolidated by use of mechanical means such as the DTS. Care must be taken on the settings (frequency, down-pressure, stabilizer speed) since these can significantly affect the resulting levels of consolidation achievable by the machine. Typically the stabilizers do not compact the ballast shoulders and for increased resistance shoulder compaction may also be useful. DTS applications should be periodically calibrated against the STPT measurements to verify their effectiveness and ensure correct settings and reliability. Coupling site-specific information with these measurements, CWR-SAFE could be exercised to evaluate buckling vulnerability at high temperatures.

Neutral Temperature

Monitoring Hardware

Monitoring CWR neutral temperature is a difficult task since at present no practical measurement capabilities which can evaluate the absolute rail force in a nondestructive manner exist, although significant research has been carried out for its development. Research to date, however, has resulted in a methodology which can be implemented as an interim solution. As stated earlier, one identifies a threshold lower limit (such as 60 °F or 70 °F) for the neutral temperature for the assurance of buckling safety. The monitoring methodology will be used to determine locations where the neutral temperature can fall below the threshold limits. The following paragraphs will discuss methods to control the neutral temperature above the threshold limit.

The authors conceptualized the Rail Uplift Device (RUD) in 1987 to determine the absolute force in the rail by correlating its deflection under an applied vertical load to the tension or compression in the rail. The rail fasteners are removed over a given length to provide free rail bending. Based on the principle, the AAR's Track Loading Vehicle was used to successfully map the neutral temperature over some sections of track on the BNSF railroad [15]. In general, it is possible to design a portable device to lift the rail with fasteners removed and measure the rail deflection under a given vertical force level, thus correlating rail deflection level to the rail longitudinal force in the rail.

Since RUD can be cumbersome for systematic applications, its use can be limited to a one time application to determine the absolute rail force. A strain gage can be fixed at the location and a reference can be provided to the strain gage using the rail force measured by the RUD. This will avoid rail cutting, which is the usual way of providing a zero reference to the strain gage. Thus, the strain gage usage in conjunction with RUD will eliminate the undesirable rail cutting. With the initial reference from RUD, the strain gage can provide all subsequent data at the location relatively easily and quickly and for a sufficiently long time if protected properly.

Strain gages can also be applied to monitor neutral temperature changes without knowing the absolute neutral temperature values. Strain gages can be installed and zeroed at any temperature, T_i. If a reading is taken at a temperature, T_R, then the neutral temperature change can be calculated using the formula

$$\Delta T_N = e_R - \alpha(T_R - T_i) \qquad \text{(Equation 6-1)}$$

where e_R is the strain gage reading at the rail temperature T_R. If ΔT_N indicates a drop of more than a prescribed value (30 °F, for example), then the track segment's rail may need to be cut to determine the correct absolute T_N and restress as required. Alternatively, a RUD type device can be used if rail cutting must be avoided.

Monitoring Locations

The following critical locations should be identified and monitored as a priority:

- Locations with a buckling history
- Recently maintained locations (surfaced or realigned)
- Winter repaired and readjusted rail segments
- Locations consistent with heavy braking and traction forces
- Locations with excessive rail movement or bunching
- Locations with sudden changes in longitudinal resistance such as bridges and turnouts
- and other locations as described in Section 2.2 of Appendix D

Monitoring Methodology

Monitoring the exact rail force and neutral temperature is an involved process. However, the concept of minimum required neutral temperature is very useful for monitoring as it simplifies the problem to a large extent.

Appendix D shows that one measurement in a 100 foot section is adequate to determine the neutral temperature within ±5 °F maximum error and a confidence limit of 90 percent. Each of the critical locations is divided into a number of cells of 100 foot in length. One or two cells should be tested for the minimum neutral temperature values (say 60 °F or 70 °F) for the critical segment. Testing is done through either a RUD (if available) or through the strain-gage/rail cut method. If a minimum condition is found, the necessary restressing (or other required maintenance action) can be performed.

Other interim monitoring schemes include the selective monitoring of rail break repairs which require a rail-plug insertion. These plugs can be pre-instrumented with strain gage coupons (with the initial zeroes established), and once the welds are made, the new (or corrected) neutral temperature is automatically determined. In such fashion, more effective control on neutral temperature could be attained.

Measures to Control

Control of neutral temperature is important at the time of installation of new CWR, when re-welding rails after destressing and restressing, and during rail fracture repairs requiring a new plug insertion. Neutral temperature control may also be required in conjunction with maintenance operations, such as curve realignments and surfacing operations involving large lifts.

1. CWR Destressing

It is a practice in the railroad industry to destress the rail by cutting whenever the heat patrols observe the rail becoming wavy and tight in the fasteners on summer days. This operation will relieve the compressive load and is usually followed up by re-welding the rail after cutting some rail out and removing anchors and fasteners on either side of the cut locations. Research has identified the following relevant technical issues and concerns:

- The correct determination of the lengths of rail to unfasten
- The correct determination of the amount of rail to cut out so that an adequate amount of gap for welding (usually 1 in.) exists for the target neutral temperature readjustment

Industry practices on these issues are not well defined and need rational guidelines which can be based on the analytical and experimental work performed in [16]. When a rail in tension or compression is cut, the rail force is redistributed from zero at the cut location to a steady state value at some distance (referred to as the influence zone). This zone can be as long as 1000 ft on either side of the cut depending on the precut rail force level and the rail/tie longitudinal resistance. The unfastening or de-anchored length can be computed according to the formula developed in [16]. If the actual de-anchored length is smaller than that calculated residual stresses would remain after re-welding resulting in a reduced neutral temperature at these locations.

Once the correct de-anchored length is prescribed, the required gap size (or rail plug length) can be made for the required target neutral temperature setting.

2. Restressing

To restress the CWR to the desired neutral temperature, the gap after destressing must be adjusted to 1 in before welding when the rail temperature reaches the neutral temperature. If the rail temperature is lower than the desired neutral temperature due to unfavorable conditions (such as cold ambient temperatures), the use of artificial heating or mechanical tensors are typically used to induce the required force level to produce the intended neutral temperature. In the case of solar or artificial heating, the gap adjustment can be made by additional cuts on the rail ends or more fasteners can be removed to obtain the correct gap size. If the initial gap is not optimal, the railroads may weld at a lower temperature, resulting in a lower neutral temperature, or they may wait for a higher temperature to obtain the 1 in. gap for welding.

The gap size is very critical when hydraulic tensors are deployed to set the neutral temperature. In the case of solar or artificial heating, the rail temperature can be readily measured before welding. In the case of tensors, the only guiding factor is the change in the gap size under the application of tensors, which is correlated to the desired neutral temperature by look-up tables. Currently, these look-up tables assume that the rails at the junction of the unfastened and fastened zones are fully constrained against longitudinal movement under the tensor application, especially at large load levels. This may be an incorrect assumption and can contribute significant errors on the nonconservative side in the resulting neutral temperature.

As an illustrative example, consider the situation when one applies a tensor on 136# rail at 40 °F with a 325 foot de-anchored section on either side of the gap. Assume that the longitudinal resistance, f_0, in fasteners is 20 lb/in/rail, and the required neutral temperature is 100 °F. In the current railroad practice, the required gap is:

$$\Delta_1(\text{current practice}) = 1 + 2L_d\alpha\Delta T = 4'' \qquad \text{(Equation 6-2)}$$

Then according to [16], if the ends are moving with the tensor-applied force, the correct gap is given by:

$$\Delta_2 \text{ (analysis)} = \Delta_1 + [AE(\alpha\Delta T)^2/f_0] = 6.95'' \qquad \text{(Equation 6-3)}$$

The gap of 4″ results in a neutral temperature of only 70 °F, incurring an error of 30 °F when using existing guidelines.

Therefore, the present railroad practices need improved guidelines on gap adjustment for installing CWR at the correct neutral temperature using tensors. The current gap size specifications should be increased by the correction term of $[AE(\alpha\Delta T)^2/f_0]$. Since this involves the longitudinal resistance f_0 and the de-anchored length L_d, which must be evaluated specifically for each site, a black box device, such as the one discussed in Appendix E currently under development, will be useful to the industry. In the interim, current best estimate guidelines for improved destressing and rail break repairs based on recent tests and analyses is provided in [28].

3. Increased Longitudinal Resistance

The fastener longitudinal resistance is an important parameter to be controlled to minimize rail longitudinal movement through the fasteners and thus retain the high neutral temperature set when the CWR is installed or destressed. Train braking and traction forces and unequal diurnal rail heating (sun and shade) resulting in unequal thermal loads tend to move the rail, which may alter the neutral temperature.

To increase the longitudinal resistance in wood tie track with cut spikes, it is important to have effective (tightly applied and maintained) rail anchors. Considerations may be given to anchoring every tie rather than every other tie, especially in areas where excessive longitudinal rail motion is evident, or using elastic fasteners.

On concrete tie tracks with elastic fasteners, the toe load between the clip and the rail base is an important factor contributing to the longitudinal resistance. The toe load reduces in service conditions, resulting in decreased resistance. The reduction in toe load can result from fatigued clips, worn pads, abraded rail seats, missing insulators, or to being incorrectly driven on.

4. Curve Breathing

Curves are usually well ballasted on the outside to withstand the rail compressive load in summer and high lateral vehicle loads. In winter, the rail tensile forces and slow speed freight train loads on the low rail can push the curve in towards its center. The result can be curve movement radially inward in winter and outward in spring and summer. Curve breathing not only weakens the lateral resistance and initiates alignment defects but also alters rail neutral temperature. To control neutral temperature changes in the curved track, generous ballasting on the inside of the curves must also be provided. Curve staking is often recommended to monitor curve breathing induced alignment variations, which are also indicators of possible neutral temperature shifts. Correcting excessive curve movement may require realignment, possibly coupled with destressing for neutral temperature readjustment.

5. Maintenance Operations

Whenever the CWR track is subjected to any maintenance operation involving moving the track, some loss may occur in its neutral temperature. This is particularly true when the track is lifted during tamping or when being mechanically lined-in-or-out. Some existing curves may need to be realigned for larger radii to permit higher speeds and sometimes to accommodate new lines. The changes in neutral temperature should be monitored at a few critical locations by means of strain gages. It is not necessary to cut the rail to monitor the change unless the change during the particular maintenance action is large (such as in excess of 20 °F). In such cases, the track must be destressed and readjusted to the correct neutral temperature.

Misalignment

Monitoring

Currently the monitoring of lateral misalignment is done by the railroads through geometry cars at various intervals to comply with Federal specifications. It would be helpful and desirable to schedule additional track geometry inspections in the summer months, especially when hot-weather kinks could become potential buckling prone conditions.

Control

It is important to keep the misalignments under the specified FRA limits for the class of track. Initial growth of misalignments can sometimes arise from track shift conditions caused by lateral loads acting on weak spots, such as weld misalignments and weak ballast locations. These initial misalignments can grow under thermal loads and train action and must be corrected through appropriate maintenance actions. After maintenance to correct track misalignments (especially after surfacing and realignment), the ballast must be consolidated to an appropriate level by either artificial consolidation methods or through slow speed traffic.

Track Buckling Mitigation

In addition to the safety implementation methodologies discussed in the previous section, the following approaches are advisable in helping to minimize the problem of track buckling.

Vehicle Dynamic Loads

Vehicle net axle dynamic loads on curves with cant deficiency may cause progressive track shift, leading to a buckling scenario, particularly in the presence of high compressive loads. Excessive unbalanced speeds must be avoided, although this is the trend in modern high-speed passenger trains to reduce travel time. Likewise, the net axle lateral-to-vertical force ratios must be low ($L/V \leq 0.4$) while the vehicle negotiates any lateral misalignment on tangent or curved tracks, as per [27]. In the similar vain, heavy traction and braking forces should be minimized, since these can also trigger buckles under hot temperature conditions.

Slow Speeds

Slow speeds are recommended whenever rail temperatures are above the T_c critical values discussed in earlier sections. Slow speeds on curves and misaligned tracks not only reduce track shift potential but also reduce the damage should a buckling event occur.

Train Handling

One important aspect of train handling is the development of procedures for the engineer to engage brake applications or implement other evasive measures when encountering a buckle in front of the locomotive caused by the preceding train. Such procedures are very difficult to devise on the account of the many variables of each particular scenario (such as speed, consist length, freight commodity, physical track location/conditions and size of the buckle in front of the train). Therefore, in lieu of such procedures, efforts should be focused on catching the buckle by the previous train causing it. Since buckles tend to occur usually toward the rear end of the consist often allowing the end vehicles to negotiate the buckles. It would be useful for the last car to contain appropriate train end devices to identify buckling occurrences and alert the locomotive engineer up front to warn the dispatcher or the following train of the potential dangerous track condition. Such devices could be based on simple accelerometer based sensors which indicate unusual vehicle lateral motion or rocking condition usually associated with traversing a large track misalignment. It is not conceptually difficult to design and implement such devices.

Summary

- CWR buckling safety limits can be derived directly by exercising the CWR-BUCKLE program with the appropriate input parameters. Alternatively, the safety limits can be assembled in the form of charts which show the relationship between the track parameters (lateral resistance, neutral temperature) and the allowable rail temperatures for safe vehicle operations with regard to lateral buckling. The relationship between lateral resistance and neutral temperature can be uncoupled in the form of minimum required values. This approach has certain advantages, including simplicity and ease of implementation without the need to determine the exact values of the parameters.

- Safety methodology using the probabilistic method has also been developed and presented. This method can provide more flexibility to the industry than offered by the deterministic method because it provides risk-based approaches to buckling prevention (such as stipulating the overall permissible number of buckles per annum in a given railroad territory). The method provides a global approach to safety specification and is complementary to the deterministic method, which is more suitable for specific locations and for day-to-day operations.

- The deterministic theory's safety limits were validated in full-scale dynamic buckling tests. The probabilistic approach needs validation, which may be accomplished through case studies in collaboration with the industry.

- Methods of safety limit implementation using the deterministic approach are presented. Monitoring and controlling the lateral resistance at or above the desired value are discussed. Estimating the resistance using the empirical equations inherent in CWR-INDY and using STPT equipment for direct measurement are available methods for this purpose. Control of lateral resistance can be achieved by providing a full ballast section and ties with good bottom friction characteristics and effective consolidation after maintenance.

- Neutral temperature monitoring and control are also important in CWR safety assurance. At present, monitoring can be done primarily with strain gages. The reference value required by the strain gage can be obtained by rail cutting or alternatively by a RUD, as described in this chapter. Control of neutral temperature may be accomplished through appropriate destressing and restressing operations. This requires knowledge of the proper unfastening or de-anchoring length on either side of the rail cut or fracture and the proper gap size for adjustment. Effective anchoring and increased longitudinal resistance of fasteners can also help maintaining desired neutral temperatures in service. Ample ballast on the inside and outside of curves can control curve breathing, and then the neutral temperature changes. Control of neutral temperature will also be necessary during track realignment and some maintenance operations.

7. Conclusions and Recommendations

The following conclusions are based on the technical studies presented in this report along with recommendations for further research.

Conclusions

- A critical mechanism in CWR track lateral buckling under thermal loads is the rail bending wave (uplift) between the trucks of a car. The uplift wave is controlled by several parameters, including, track vertical stiffness, truck central spacing, tie-ballast friction coefficient, and axle loads. Buckling under a car is identified as the more frequent occurrence as opposed to the static case (i.e., without vehicle influences).

- Buckling can be explosive or progressive. Explosive buckling (sometimes referred to as snap-through) is characterized by equilibrium jumps between the two critical temperatures (identified as T_{Bmax} and T_{Bmin}) on the buckling response curve that defines the relationship between temperature increase and lateral displacement. At the upper critical temperature, T_{Bmax}, the track buckles out automatically without any external energy. In contrast, to precipitate buckling at the lower critical temperature, some finite energy supplied by moving vehicles is required. Progressive buckling, sometimes referred to as a slow-buckle, is manifested by increasing lateral deflections with increasing temperatures when no distinct T_{Bmax} and T_{Bmin} exist. This usually occurs for weak track conditions.

- A dynamic buckling theory was formulated on the basis of the uplift mechanism taking into account large deflections and nonlinearities in resistance parameters. The theory was validated by full-scale field tests which simulated explosive and progressive buckling characteristics under vehicles. The theory and tests identified critical track and vehicle parameters that govern track buckling, and specialized measurement techniques were developed for their determination.

- The principal track parameters that influence CWR buckling include track lateral and torsional resistances, misalignments, CWR neutral temperature, rail sectional properties, tie-ballast friction characteristics, and track foundation stiffness. The knowledge and measurement of these parameters is important, thus this report presents an analysis and measurement techniques for them.

- The dynamic buckling theory has been divided into two groups. The first uses the theory in a deterministic manner, whereas the second relies on statistical principles. Computer codes have been developed for the deterministic and the probabilistic approaches. The deterministic codes are CWR-BUCKLE and CWR-INDY, which, for a given set of parameters, can ascertain whether or not the track will buckle out at a given rail temperature. The probabilistic approach, which evaluates the probability of buckling as a percentage between 0 (no buckling) and 100 (buckling) for a given scenario, is coded in CWR-RISK. All three modules are incorporated in a master code, CWR-SAFE, which operates in Windows environment on a PC.

- CWR-BUCKLE requires measured input parameters (such as the nonlinear lateral resistance characteristic defined in terms of peak and limit resistance values). The program output includes buckling response data which can be graphically represented, upper and lower critical temperatures (T_{Bmax}, T_{Bmin}), the safe allowable rail temperature increase, and the margin of safety against buckling. The program contains a track quality based safety criterion to determine allowable temperatures and buckling safety margin.

- CWR-INDY needs only physically describable input parameters, such as the type of track, ballast type, ballast crib level and shoulder width, and consolidation levels. These are used in the program to deduce the required scientific parameters quantitatively through the built-in empirical equations in the code. The output of CWR-INDY is the allowable temperature for buckling (or the buckling strength) and the margin of safety. The output also includes numerical value of the peak lateral resistance. CWR-INDY is intended for quick on-the-spot industry use for buckling strength evaluations.

- CWR-RISK works with inputs described through the statistical distributions of the three key parameters of lateral resistance, misalignment amplitude, and neutral temperature. The output of this code is the buckling probability expressed as a function of the rail temperature. From this output, critical temperatures for buckling prevention can be determined, and the expected annual number of buckles in a territory can be evaluated.

- The parametric study generated using the CWR-BUCKLE revealed information of practical significance on CWR buckling strength in terms of key variables and their sensitivity on the buckling response characteristics. For example, the peak value of lateral resistance has a significant influence on the upper temperature, T_{Bmax}, as has the lateral misalignment. The lower buckling temperature, T_{Bmin}, is influenced by the limiting value of the lateral resistance, torsional resistance, and lateral misalignment. Both temperatures are strongly influenced by track curvature. Vehicle truck center spacing and axle loads also impact buckling temperatures through their influence on the dynamic (uplift) lateral resistance.

- The parametric study based on the probabilistic approach identifies a critical temperature, T_c, below which the buckling probability is zero. Above this T_c value, the probability is found to increase rapidly with rail temperature, and thus, T_c is identified as the temperature for the deployment of preventative measures such as slow orders. The results of the probabilistic approach can also be used to define a limiting rail temperature, T_L, beyond which vehicle operations should be stopped or carried out at low speeds. This T_L can be determined by choosing a number of acceptable buckles per year; thus the railroads can control future annual number of buckles by adhering to T_c as the safe limit and curtailing operations (such as slow-orders) beyond it up to T_L.

- Controlling T_c and the number of buckles by an appropriate choice of T_L will also require controlling the low-end values of the distributions of lateral resistance, neutral temperature, and the high-end values of lateral misalignments. Numerical examples indicate that controlling the neutral temperature's low-end values is particularly effective for buckling prevention.

- Based on a risk methodology, a speed reduction formula has been devised relating vehicle speed (as a percent of maximum line speed) to the rail temperature, $T_R > T_c$. The formula produces zero speed when $T = T_L$, the limiting rail temperature for vehicle operations. Applying speed restrictions between T_c and T_L becomes effective in limiting

111

the annual number of buckles to accepted values, as well as reducing the consequences of the buckling induced derailments.

- Safety limits are formulated on the basis of the deterministic approach. The limits were validated by full-scale tests with moving vehicles on heated CWR test segments. For the safety implementation of these limits, the approach is recommended that stipulates the adherence to established minima for lateral resistance and neutral temperature for a given FRA class of track (i.e., for a stipulated maximum allowable misalignment). Methods to verify these minimum values include the STPT device or CWR-INDY for lateral resistance prediction, and the RUD and strain gage for neutral temperature.

- Using the data collected previously, the required sample rates for the measurements of lateral resistance and neutral temperature have been estimated. In the case of lateral resistance, an error of 15 percent is tolerable for the resistance range of practical interest (1500 to 2000 lb). This will not contribute to an error of more than 5 °F in the allowable rail temperature for safe operations of vehicles on CWR tracks. To evaluate the resistance within this maximum permissible error and for a confidence limit of 90 percent, one sample in a 50 ft section is adequate for concrete tie track, and two samples over a 50-foot section for wood tie tracks. Similarly, for the case of neutral temperature measurement, for a tolerable error of 5 °F, one measurement over a 100-foot section seems adequate for the measurement with 90 percent confidence limit.

Recommendations

- The probabilistic approach and its application to risk-based buckling prevention practices should be demonstrated on cooperating railroads. This would require testing for the statistical variations in lateral resistance, neutral temperature, and lateral alignment within a characteristic or representative line segment. The data would be used in CWR-RISK to develop safer operating temperature regimes for CWR, as well as to indicate potential cost-effective maintenance strategies.

- Implement the Rail Uplift technique onto a moving platform, such as on a hi-railer or a dedicated rail car, to facilitate an easy and quick measurement of the minimum required neutral temperature.

- Develop a simplified STPT and procedure for a quick peak resistance measurement (i.e., mobilizing the test tie only through a small 0.2–0.4 in deflection by eliminating complex hydraulics and simplifying data acquisition).

- A rail destressing force indicator can be developed to more effectively install and destress CWR by setting the neutral temperature more correctly. Such an instrument can also be used to calibrate the hydraulic rail tensors currently in maintenance use to minimize possible errors in their present application.

- Continue testing revenue service tracks for neutral temperature determination to better identify buckling prone conditions and to provide data toward establishing better neutral temperature management practices, including more effective destressing and repairing broken rail repair procedures.

112

- Using STPT, further quantify the sampling rates established in Appendix D. Further, correlate with field data the empirical equations for lateral resistance in CWR-INDY. It is also important to continue lateral resistance mapping on revenue lines to better define statistical variations for CWR-RISK applications, as well as for data to further develop and validate CWR-INDY's lateral resistance predictor.

- Evaluate the risk-based slow order methodology presented here against railroad experience to develop more extensive applications strategies, including the promulgation of risk-index based approaches to buckling prevention.

8. References

1. U.S. DOT/Federal Railroad Administrations Accident/Incident Bulletins 149-165 (1980-1996); and Annual Reports on Safety Statistics Annual Reports (1997-2003).

2. Samavedam, G., "Buckling and Post Buckling Analyses of CWR in the Lateral Plane," British Railways Board, R&D Division, Technical Note, TN TS 34, 1979.

3. Kish, A., and G. Samavedam, "Analysis of Thermal Buckling Tests on United States Railroads," DOT/FRA/ORD-82/45, 1982.

4. Kish, A., G. Samavedam, and D. Jeong, "Influence of Vehicle Induced Loads on the Lateral Stability of CWR Track," DOT/FRA/ORD-85/03, 1985.

5. Samavedam, G., A. Kish, and D. Jeong, "Experimental Investigation of Dynamic Buckling of CWR Tracks," DOT/FRA/ORD-86/07, November 1986.

6. Kish, A., and G. Samavedam, "Analyses of Phase III Dynamic Buckling Tests," DOT/FRA/ORD-89/08, February 1990.

7. Kish, A., and G. Samavedam, "Dynamic Buckling Test Analyses of a High Degree CWR Track," DOT/FRA/ORD-90/13, 1991.

8. Kish, A., and G. Samavedam, "Dynamic Buckling of Continuous Welded Rail Track: Theory, Tests, and Safety Concepts," Transportation Research Record, 1289, Proceedings of Conference on Lateral Track Stability, 1991.

9. Samavedam, G., A. Purple, A. Kish, and J. Schoengart, "Parametric Analysis and Safety Concepts of CWR Track Buckling," Final Report, DOT/FRA/ORD-93/26, December 1993.

10. Samavedam, G., A. Kanaan, J. Pietrak, A. Kish, and A. Sluz, "Wood Tie Track Resistance Characterization and Correlations Study," Final Report, DOT/FRA/ORD-94/07, January 1995.

11. Kish, A., D.W. Clark, and W. Thompson, "Recent Investigations on the Lateral Stability of Wood and Concrete Tie Tracks," AREA Bulletin 752, October 1995, pp 248-265.

12. Samavedam, G., A. Kish, and D. Jeong, "The Neutral Temperature Variation of Continuous Welded Rails," AREA Bulletin 712, 1987.

13. Kish, A., and G. Samavedam, "Risk Analysis Based CWR Track Buckling Safety Evaluations," Proceedings of Conference on "Innovations in the Design and Assessment of Rail Track," December 1999, Delft University of Technology, The Netherlands.

14. Kish, A., and G. Samavedam, "Longitudinal Force Measurement in Continuous Welded Rail from Beam Column Deflection Response," AREA Bulletin 712, Vol. 88, 1987.

15. Kish, A., S. Kalay, A. Hazell, J. Schoengart, and G. Samavedam, "Rail Longitudinal Force Measurement Evaluation Studies Using the Track Loading Vehicle," Bulletin 742, American Railway Engineering Association, Bulletin No. 742, 1993.

16. Samavedam, G., "Investigation on CWR Longitudinal Restraint Behavior in Winter Rail Break and Summer Destressing Operations," DOT/FRA/ORD-97/01, 1997.

17. ERRI Report, Samavedam, G., "Theory of CWR Track Stability," D202/RP3, Utrecht, February 1995.

18. Samavedam, G., et. al., "Analysis of Track Shift Under High-Speed Vehicle-Track Interaction," Technical Report DOT/FRA/ORD-97/02, 1997.

19. Thomson, D., G. Samavedam, W. Mui, and A. Kish, "Field Testing of High Degree Revenue Service Track for Buckling Safety Assessment," Technical Report, DOT/FRA/ORD-92/02, November 1993.

20. Union of International Railways (UIC) Code 720R, "Laying and Maintenance of CWR Track," 2nd Edition, December 2002.

21. Klaren, J.W., and J.C. Loach, "Lateral Stability of Rails, Especially of Long Welded Rails," Question D14, ORE, Utrecht, 1965.

22. Sussmann, T., A. Kish, and M. Trosino, "Influence of Track Maintenance on the Lateral Resistance of Concrete Tie Track," Transportation Research Board Record, 1825, Paper No. 03-3694.

23. Samavedam, G., and A. Kish, "Track Lateral Shift Model Development and Test Validation," DOT/VNTSC/FRA Report, 2002.

24. Harnett, D., "Statistical Methods, Third Edition," The Philippines: Addison Wesley Publishing Company, Inc., 1982.

25. Kristoff, S., "Track Lateral Strength Measurements at Union Pacific Railroad Sites," Foster-Miller Report prepared for Union Pacific Railroad, 2001.

26. Kish, A., T. Sussman, and M. Trosino, "Effects of Maintenance Operations on Track Buckling Potential," International Heavy Haul Association Technical Conference, May 2003, Dallas, Texas.

27. Kish, A., G. Samavedam, and D. Wormley, "New Track Shift Limits for High-Speed Rail Applications," World Congress for Railway Research (WCRR 2001), November 25-29 2001, Cologne, Germany.

28. Kish, A. and G. Samavedam, "Improvements in CWR Destressing for Better Management of Rail Neutral Temperature," Transportation Research Board 2005 Annual Conference, January 9-12, 2005.

Appendix A.
Mathematical Formulation of CWR-Buckle

A.1 Buckling Theory

The equation to analyze the lateral stability of continuous welded rail (CWR) track can be derived by applying the principle of minimum potential energy under the assumptions of Chapter 2. Through the use of variational calculus, the equations of equilibrium are presented in the form of two highly nonlinear differential equations for tangent and curved tracks. These differential equations become solvable when the infinite track domain is divided into two regions: (1) a buckled zone where longitudinal displacement is neglected and (2) an adjoining zone that extends to infinity where lateral displacement is neglected (Figure A-1). The equations for the two regions are coupled through another equation, which yields the temperature. For an assumed buckled length, 2L, the temperature increase, the lateral track deflection, and the compressive force in the rails can be calculated through the solution of the following differential equations.

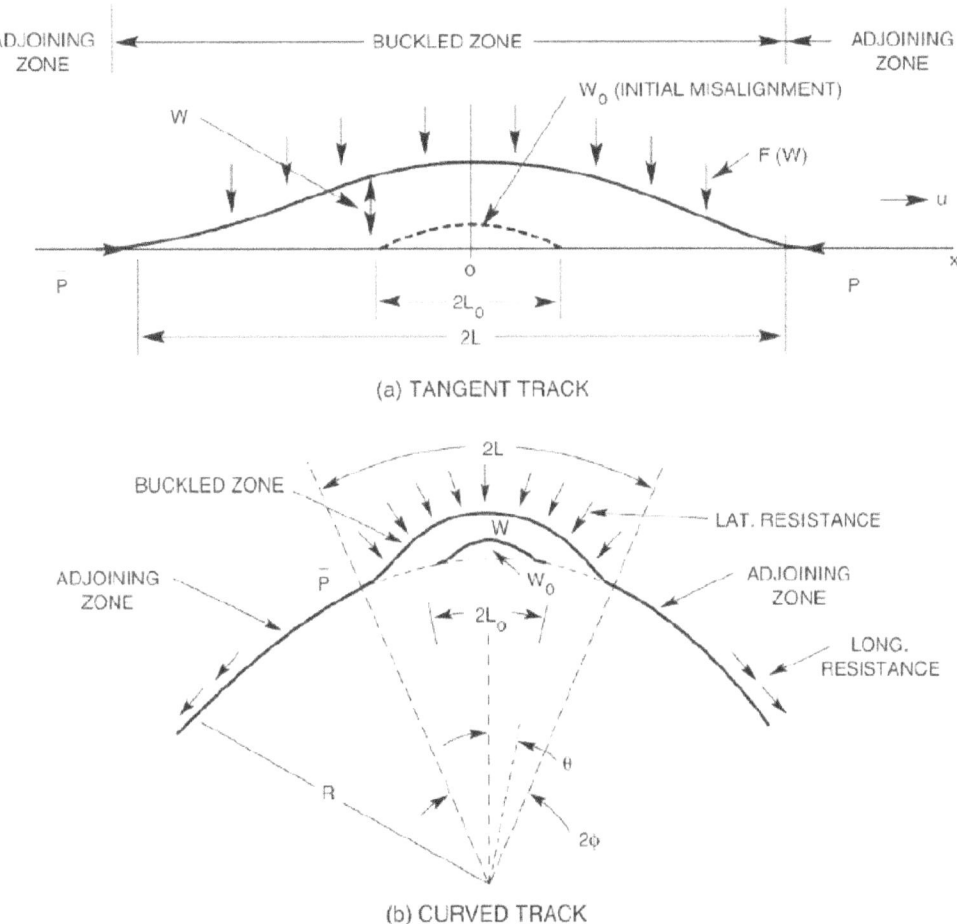

(a) TANGENT TRACK

(b) CURVED TRACK

Figure A-1. Geometry and Coordinates

A.2 Differential Equations and Solution

The basic differential equations are formulated using the large deflection theory and principles of variational calculus as in [2].

Buckling Zone ($0 \leq x \leq L$): Tangent Track

The differential equation is

$$EI_{zz}\frac{d^4 w}{dx^4} + \left(\bar{P} - \tau_o\right)\frac{d^2 w}{dx^2} = -F[w(x)] - \bar{P}\frac{d^2 w_o}{dx^2} \tag{A-1}$$

where

E	= the modulus of elasticity,
I_{zz},	= the area moment of two rails for lateral bending (i.e., about the vertical axis),
$\bar{P}$	= the longitudinal force in the rails,
$F[w(x)]$	= the lateral resistance distribution function,
τ_o	= the linear torsional stiffness of fasteners (both the lateral resistance and the torsional stiffness are expressed here per unit track length),
w_o	= the initial imperfection distribution, and
w	= the deflection in the lateral direction.

The solution of this differential equation can be expressed in terms of an infinite trigonometric series, assuming the initial misalignment and the buckling mode are symmetric.

$$w(x) = \sum_{m=1,3,5,\ldots}^{\infty} A_m \cos\left(\frac{m\pi x}{2L}\right) \tag{A-2}$$

$$\frac{d^2 w_o}{dx^2} = \sum_{m=1,3,5,\ldots}^{\infty} b_m \cos\left(\frac{m\pi x}{2L}\right) \tag{A-3}$$

$$F[w(x)] = \sum_{m=1,3,5,\ldots}^{\infty} a_m \cos\left(\frac{m\pi x}{2L}\right) \tag{A-4}$$

Where

$$A_m = \frac{-\left(a_m + \bar{P}b_m\right)}{EI_{zz}\left(\frac{m\pi}{2L}\right)^4 - \left(\bar{P} - \tau_o\right)\left(\frac{m\pi}{2L}\right)^2} \tag{A-5}$$

The foregoing expressions satisfy the requirements of zero deflection and moment at the ends of the buckled zone. In addition, the zero slope condition will be satisfied by stipulating

$$\sum_{m=1,3,5,}^{\infty} mA_m \sin\left(\frac{m\pi}{2}\right) = 0 \qquad \text{(A-6)}$$

The foregoing equation which is solved by an iteration scheme, gives the relationship between the assumed buckling length 2L, and the compressive load $\overline{P}$.

Adjoining Zone ($|x| \geq L$)

The longitudinal resistance is assumed proportional to longitudinal displacement. The governing differential equation in the adjoining zone ($x > L$) is derived from equilibrium considerations in the longitudinal direction. Thus, if proportional longitudinal resistance is assumed, then the differential equation in the adjoining zone is

$$AE\frac{d^2U}{dx^2} = k_f U \qquad \text{(A-7)}$$

where

    A   = the cross sectional area of two rails,
    E   = the modulus of elasticity,
    k_f = the slope of the longitudinal resistance versus longitudinal displacement curve
             (i.e., longitudinal resistance stiffness),
    U   = the longitudinal displacement.

The solution to this equation is

$$U(x) = C_1 e^{\psi x} + C_2 e^{-\psi x} \qquad \text{(A-8)}$$

where

$$\psi^2 = \frac{k_f}{AE}$$

However, the solution must be bounded for very large values of x. Then, $U = U' = 0$ in., the limit as x approaches infinity, and $C_1 = 0$. After differentiation,

$$U'(x) = -\psi C_2 e^{-\psi x}$$

or

$$U' = -\psi U \qquad \text{(A-9)}$$

Buckling Zone ($0 \leq \theta \, \Box \, \phi$): Curved Track

The governing differential equation in the buckled zone ($0 \leq \theta \, \Box \, \phi$) for curved track is

$$\frac{EI_{zz}}{R^4}\frac{d^4w}{d\theta^4} + \left(\frac{\overline{P} - \tau_o}{R^2}\right)\frac{d^2w}{d\theta^2} = -F[w(\theta)] + \frac{\overline{P}}{R} - \frac{\overline{P}}{R^2}\frac{d^2w_o}{d\theta^2} \qquad \text{(A-10)}$$

A-3

As in the tangent case, the solution to this equation can be expressed in terms of an infinite trigonometric series

$$w(\theta) = \sum_{m=1,3,5,...}^{\infty} A_m \cos\left(\frac{m\pi\theta}{2\phi}\right)$$

(A-11)

$$\frac{\overline{P}}{R^2}\frac{d^2 w_o}{d\theta^2} = \sum_{m=1,3,5,}^{\infty} b_m \cos\left(\frac{m\pi\theta}{2\phi}\right)$$

(A-12)

$$F[w(x)] = \sum_{m=1,3,5,}^{\infty} a_m \cos\left(\frac{m\pi\theta}{2\phi}\right)$$

(A-13)

$$\frac{\overline{P}}{R} = \sum_{m=1,3,5,...}^{\infty} \frac{\overline{P}}{R} c_m \cos\left(\frac{m\pi\theta}{2\phi}\right)$$

(A-14)

where

$$A_m = \frac{-\left(a_m - \dfrac{\overline{P}}{R}c_m + \dfrac{\overline{P}}{R^2}b_m\right)}{\dfrac{EI_{zz}}{R^4}\left(\dfrac{m\pi}{2\theta}\right)^4 - \left(\dfrac{\overline{P}-\tau_o}{R^2}\right)\left(\dfrac{m\pi}{2\theta}\right)^2}$$

(A-15)

The foregoing equations satisfy zero deflection and curvature requirements at the ends of the buckled zone. An equation similar to that of A-6 satisfying the zero slope condition relates the compressive force, $\overline{P}$, to the assumed buckling length, 2L.

Adjoining Zone: Curved Track

The differential equation of longitudinal equilibrium that applies to the adjoining zone ($\theta > \phi$) for the curved track case, again assuming proportional longitudinal resistance, is

$$\frac{AE}{R^2}\frac{d^2 U}{d\theta^2} = k_f U$$

(A-16)

Recall that $L = R\phi$ and $x = R\theta$. Thus, the general solution to this equation is

$$U(\theta) = C_3 e^{R\psi\theta} + C_4 e^{-R\psi\theta}$$

(A-17)

where

$$\psi^2 = \frac{k_f}{AE}$$

The function F in A-1 and A-10 represents the lateral resistance. F can be expressed as:

$$F = F_P F(w)$$

(A-18)

where F_P is the peak value, which can be a function of x, and $F(w)$ is the functional dependency on the lateral displacement w.

For the case where vehicle loading is present, the peak resistance is a function of the longitudinal distance along the track

$$F_{P(dynamic)} = \begin{cases} \left[F_P - \mu_f Q\right] & \text{for uplift} \\ \left[F_P + \mu_f R_v(x)\right] & \text{otherwise} \end{cases}$$

(A-19)

where

F_P = the peak value of static lateral resistance,
μ_f = the tie-ballast coefficient of friction,
Q = the self weight of the track,
$R_v(x)$ = the ballast vertical reaction to the vehicle wheel loads on the track.

The vertical reaction can be calculated from classical beam on elastic foundation theory as shown later. Uplift occurs when the sum of the vertical deflection and the self weight of the track is less than zero or mathematically when $[Q + R_v(x)] < 0$.

The Fourier coefficient that accounts for the effect of lateral resistance on the track, a_m, is derived from the following integral

For tangent track

$$a_m = \frac{2}{L} \int_o^L F[w(x)] \cos\left(\frac{m\pi x}{2L}\right) dx$$

(A-20)

For curved track

$$a_m = \frac{2}{L} \int_o^\phi F[w(\theta)] \cos\left(\frac{m\pi\theta}{2\phi}\right) d\theta$$

(A-21)

The integrals are evaluated using Filon's integration scheme. If the lateral resistance function $F[w(x)]$ is a constant value, F_o, then this integral can be evaluated in closed form

$$a_m = \frac{4F_o}{m\pi} \sin\left(\frac{m\pi}{2}\right)$$

(A-22)

The Fourier coefficient that accounts for the effect of initial imperfection in the track, b_m, is derived from the following integral

For tangent track

$$b_m = \frac{2}{L} \int_o^{L'} \frac{d^2 w_o}{dx^2} \cos\left(\frac{m\pi x}{2L}\right) dx$$

(A-23)

where

$$L' = \begin{cases} L & \text{if } L \le L_o \\ L_o & \text{if } L > L_o \end{cases}$$

and

L_0 = misalignment half-wavelength

For curved track

$$b_m = \frac{2}{\phi} \int_o^{\phi'} \frac{d^2 w_o}{d\theta^2} \cos\left(\frac{m\pi\theta}{2\phi}\right) d\theta \tag{A-24}$$

where

$$\phi' = \begin{Bmatrix} \phi \ \text{if}\ \phi \leq \phi_o \\ \phi_o \ \text{if}\ \phi > \phi_o \end{Bmatrix}$$

<u>Misalignment</u>

The initial imperfection shape is assumed as a fourth degree polynomial

$$w_o(x) = \delta_o \left[1 - 2\left(\frac{x}{L_o}\right)^2 + \left(\frac{x}{L_o}\right)^4 \right] \tag{A-25}$$

where

δ_o = the offset or the misalignment amplitude,

L_0 = half the wave length over which the misalignment occurs.

For the imperfection shape shown above, evaluation of A-23 results in the following

For $L \leq L_0$

$$b_m = -\frac{16\delta_o}{m\pi L_o^2} \left\{ 1 - 3\left(\frac{L}{L_o}\right)^2 \left[1 - 2\left(\frac{2}{m\pi}\right)^2 \right] \right\} \sin\left(\frac{m\pi}{2}\right) \tag{A-26}$$

For $L > L_0$

$$b_m = -\frac{16\delta_o}{m\pi L_o^2} \left\{ -6\left(\frac{L}{L_o}\right)\left(\frac{2}{m\pi}\right) \cos\left(\frac{m\pi L_o}{2L}\right) + 2\left[-1 + 3\left(\frac{2L}{m\pi L_o}\right)^2 \right] \sin\left(\frac{m\pi L_o}{2L}\right) \right\} \tag{A-27}$$

For $\phi \ \square \leq \square \ \phi_o$

$$\frac{b_m}{R^2} = -\frac{16\delta_o}{m\pi L_o^2} \left\{ 1 - 3\left(\frac{L}{L_o}\right)^2 \left[1 - 2\left(\frac{2}{m\pi}\right)^2 \right] \right\} \sin\left(\frac{m\pi}{2}\right) \tag{A-28}$$

For $\phi \ \square > \square \ \phi_o$

A-6

$$\frac{b_m}{R^2} = -\frac{16\delta_o}{m\pi L_o^2}\left\{-6\left(\frac{L}{L_o}\right)\left(\frac{2}{m\pi}\right)\cos\left(\frac{m\pi L_o}{2L}\right) + 2\left[-1 + 3\left(\frac{2L}{m\pi L_o}\right)^2\right]\sin\left(\frac{m\pi L_o}{2L}\right)\right\}$$

(A-29)

$$\phi = \frac{L}{R} \text{ and } \phi_o = \frac{L_o}{R}$$

Also,

$$c_m = \frac{2}{\phi}\int_o^\phi \cos\left(\frac{m\pi\theta}{2\phi}\right)d\theta = \frac{4}{m\pi}\sin\left(\frac{m\pi}{2}\right)$$

(A-30)

Temperature Calculations

The temperature equation is derived by using continuity requirements on the longitudinal displacement between the buckled and adjoining zones. It can be shown that

$$U(L) = -\frac{\overline{P}L}{AE} - Z + \alpha\Delta TL$$

(A-31)

$$U'(L) = -\frac{\overline{P}L}{AE} + \alpha\Delta T$$

(A-32)

where L is the buckled length and Z is defined below. Using A-10

$$T = \frac{\overline{P}}{AE\alpha} + \frac{Z\psi}{\alpha(1 + \psi L)}$$

(A-33)

where

$$Z = \int_o^L \left(\frac{w'^2}{2} + w'w'_o\right)dx$$

(A-34)

The equation for Z can be rewritten, after an integration by parts

$$Z = \int_o^L \left(\frac{w'^2}{2} - ww''_o\right)dx$$

(A-35)

The expression for Z can be expressed as an infinite series by applying Fourier analysis

$$Z = \frac{L}{4}\sum_{m=1,3,5,\ldots}^\infty \left[A_m^2\left(\frac{m\pi}{2L}\right)^2 - 2A_m b_m\right]$$

(A-36)

In a similar fashion to the tangent analysis, the temperature equation for curved track is

$$T = \frac{\overline{P}}{AE\alpha} + \frac{ZR\psi}{\alpha(1 + \psi L)}$$

(A-37)

A-7

where the equation for Z can be written as (after integration by parts and application of a Fourier series)

$$ZR = \sum_{m=1,3,5,...}^{\infty} \left[\frac{2L}{m\pi R^2} A_m \sin\left(\frac{m\pi}{2}\right) + \left(\frac{m\pi}{2}\right)^2 \frac{A_m^2}{4L} - \frac{A_m b_m L}{2R^2} \right]$$

(A-38)

In the limit as R approaches infinity or as the track curvature becomes tangent, the expression for ZR approaches the expression for Z in the tangent case since b_m for the tangent case is identical to b_m/R^2 for curved. Thus, the two temperature equations also reduce to the same expression in the limiting process.

Vertical Deflection and Reaction Calculation

Quasi-static load idealization is assumed to be adequate in determining loss of resistance in the uplift region, which occurs due to the vertical track deformation under wheel loads. The differential equation for the vertical deflection v is

$$EI_{yy} v'''' + K_v v = \sum \delta_i (x - x_i) V_i + Q$$

(A-39)

where

EI_{yy} = combined flexural rigidity for the two rails in the vertical plane
K_v = track foundation stiffness (assumed constant)
V_i = vertical wheel loads
δ_i = Dirac delta functions
Q = track weight/unit length

In this equation, the effects of vertical imperfections and the compressive load in the rails are excluded for the sake of simplicity.

After solving equation A-39 under the boundary conditions at infinity, $v = v' = 0$, one can compute the distributed foundation (tie ballast) reaction $R_v(x)$ given by

$$R_v = K_v \bullet v(x)$$

(A-40)

A.3 Energy Required for Buckling

As stated earlier, the upper buckling temperature represents stability in the infinitesimal sense requiring no external energy for snap-through explosive buckling. At temperatures lower than this (but higher than the lower buckling temperature), the track can buckle out upon the application of a finite external energy. Thus the energy needed to cause buckling can be used as a measure of the degree of stability. This measure will be useful in the development of CWR safety limits.

Referring to Figure 2-3c in Chapter 2, the pre-buckling state is represented by position (1) while the post-buckling unstable branch is represented by position (2). It is assumed that, if the track can be brought into position (2), it will automatically move to position (3).

The following quantities are defined.

  V_1 = strain energy in the rails at stable equilibrium position (1)
  V_2 = strain energy in the rails at unstable equilibrium position (2)
  W = work done against resistances by moving track from position (1) to position (2)
  Ω = energy required to move track from position (1) to position (2)

By an energy balance

$$\Omega = (V_2 - V_1) + W \tag{A-41}$$

The strain energy components are given by the following integrals

$$V_1 = \frac{1}{2}\int_0^\infty \frac{P_\infty^2}{AE}dx \tag{A-42}$$

where

$$P_\infty = -AE\alpha T$$

Here, for simplicity, neglect the energy due to bending in the pre-buckling state

$$V_2 = \frac{1}{2}\int_0^\infty \frac{P^2}{AE}dx + \frac{EI_{zz}}{2}\int_0^\infty \left(\frac{d^2 w}{dx^2}\right)^2 dx \tag{A-43}$$

In the curved track case, the longitudinal force distribution becomes

$$P = \begin{cases} \overline{P} & \text{for } 0 \le \theta \le \phi \\ AE\left(\frac{1}{R}\frac{du}{d\theta} - \alpha T\right) & \text{for } \theta > \phi \end{cases} \tag{A-44}$$

The work done against the lateral and longitudinal resistances are given by the following integrals

$$W_1 = \int_0^\infty \int_0^{w(x)} F[w(x)]dw\,dx \tag{A-45}$$

$$W_2 = \int_0^\infty \int_0^{u(x)} f[u(x)]du\,dx \tag{A-46}$$

Here f is the longitudinal resistance.

Thus, the total work done against ballast resistance (lateral and longitudinal) is

$$W = W_1 + W_2 \tag{A-47}$$

The difference in strain energy is calculated from the following equation

$$V_2 - V_1 = \frac{1}{2}\int_0^\infty \frac{P^2 - P_\infty^2}{AE}dx + \frac{EI_{zz}}{2}\int_0^\infty \left(\frac{d^2 w}{dx^2}\right)^2 dx \tag{A-48}$$

A-9

This equation shows that the total strain energy is the sum of two components: one due to compressive axial force and the other due to beam bending. The evaluation of these integrals is performed with the aid of the Fourier analysis.

$$V_2 - V_1 = \frac{AE}{2}\left\{\frac{\overline{P}}{AE}\left[\frac{\overline{P}}{AE}\left(L + \frac{1}{2\psi}\right) + \frac{\alpha T}{\psi}\right] - (\alpha T)^2\left(L + \frac{3}{2\psi}\right)\right\} + \frac{EI_{zz}}{64L^3}\sum_{m=1,3,5,...}^{\infty}(m\pi)^4 A_m^2$$

(A-49)

The work done against lateral resistance can be evaluated from A-45 once the lateral resistance is expressed in terms of a mathematical function.

The work done against a linear longitudinal resistance $f = k_f u$ is given by

$$W_2 = \frac{k_f}{4\psi^3}\left(\frac{\overline{P}}{AE} - \alpha T\right)^2$$

(A-50)

A-10

Appendix B.
Techniques for Parameter Measurements

The following paragraphs describe the hardware and measurement techniques for the critical parameters developed as part of this track stability research program. Some of the measurements/technique can be greatly simplified for the purpose of routine safety assurance and implementation by the rail industry, as discussed in Chapter 6.

The hardware and measurement techniques for the following parameters are discussed.

1. Lateral Resistance
2. Torsional Stiffness
3. Longitudinal Resistance
4. Tie-Ballast Friction Coefficient
5. Rail Force and Neutral Temperature

B.1 Lateral Resistance Measurement

A number of research organizations have measured track lateral resistance in the United States and abroad. The recommended measurement scheme mobilizes only a single tie. Some of the previous techniques require lateral movement of a cut panel or the entire track section by a concentrated lateral load. In the case where only a single tie is mobilized, the resistance is directly represented by the load-deflection response of the tie and provides the spring type resistance required by analysis.

The advantages of the Single Tie Push Test (STPT) over the panel test include the following:

- STPTs yield a more fundamental and correct characteristic of the ballast resistance.
- The test is easy to set up and perform.
- The hardware is man-portable and can be used by track crew with minimal training.
- The obtained test data also provides the parameter required for other track analyses, such as lateral panel shift

The disadvantage of the STPT is the variation of the results from tie to tie. However, an arithmetic average of the individual test results is adequate for buckling analysis. Appendix D discusses the sampling rate for a statistically significant value.

B.1.1 Test Hardware

Lightweight portable devices were developed for the wood and concrete tie tracks. The devices are shown in Figure B-1 for a wood tie and Figure B-2 for a concrete tie. The device consists of a hydraulic control unit with a pump and a fixture with a hydraulic cylinder. With the spikes, rail anchors, and tie plates removed from both rails, the assembly is set to grab the test tie which is now free to move laterally under the rails. The hydraulic piston mounted on the fixture reacts with the force required to move the tie against one of the rails. Hydraulic pressure can be provided by the hand pump or by an electric pump to speed up the operation.

A pressure transducer or load cell in line with the piston and pressure gage in the control unit (as a backup) indicates the load applied. A rotary potentiometer mounted on the tie measures the displacement with respect to the stationary second rail. The load-displacement relationship is plotted using an X-Y plotter. Alternately, a modern Data Acquisition System can display this relationship on a monitor and store the data on a computer.

Figure B-1. STPT Equipment for Wood Ties

Figure B-2. STPT Equipment for Concrete Ties

B.1.2 Typical Results

A large number of track resistance tests were conducted using the STPT device at the Transportation Test Center's (TTC) test tracks in Pueblo, CO, and on a number of railroads. Figure B-3 shows typical results for relatively strong, medium, and weak wood and concrete tie tracks. In general, two salient points on the load versus deflection characteristics exist: the peak F_P, occurring at displacements on the order of 0.25-0.5 in and the limiting value, F_L, at about 3-5 in of tie displacement. The softening or drooping behavior becomes more pronounced for high F_P values; whereas for low F_P's, the resistance is practically constant with F_P being close or equal to F_L.

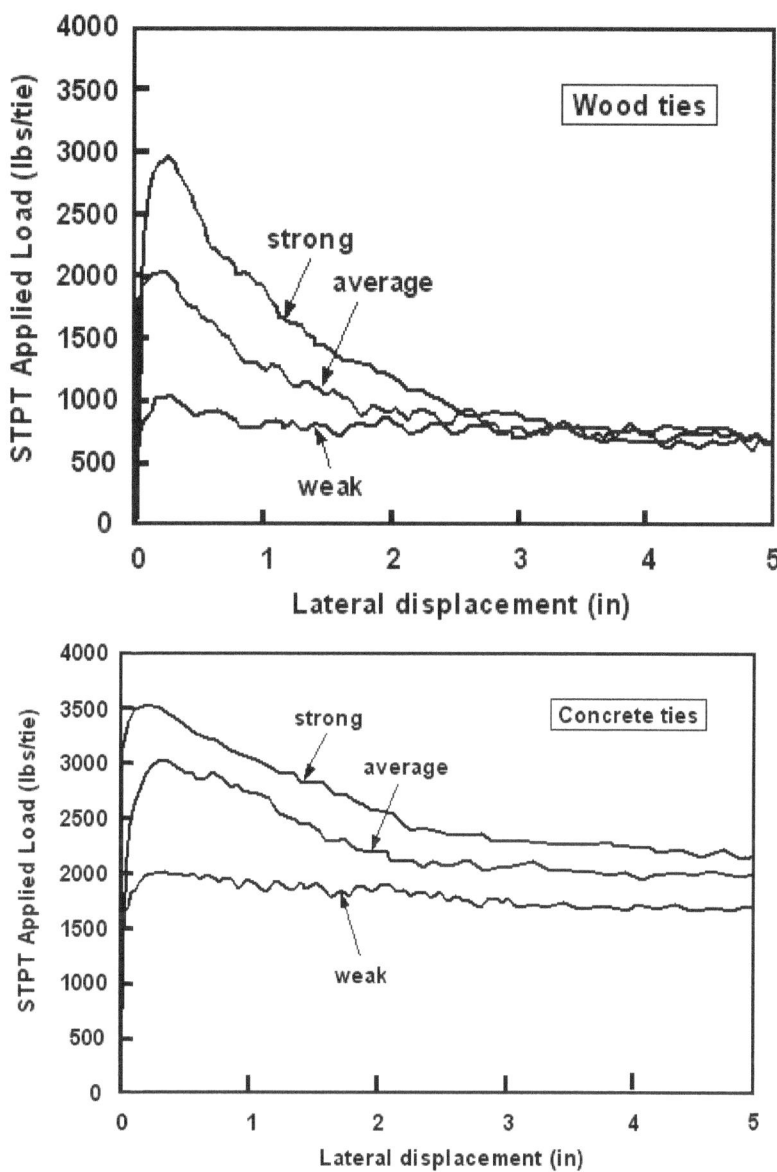

Figure B-3. Typical STPT Results for Strong, Average, and Weak Track for Wood and Concrete Ties

The most recent application of the STPT for track lateral resistance evaluation was at an Amtrak concrete tie track segment at New Carrollton, MD. The tests were to evaluate track lateral strength and stability after track maintenance and subsequent stabilization operations on Class 4 mainline track segment. The tests consisted of performing track lateral resistance measurements using the STPT, and comparing the resulting measurements after maintenance and different consolidation operations for subsequent buckling safety evaluations. Figure B-4 provides a data summary, and additional test results are available in [22]. The data was also used to conduct buckling safety analyses using CWR-RISK to evaluate buckling risk implications resulting from the maintenance actions [26].

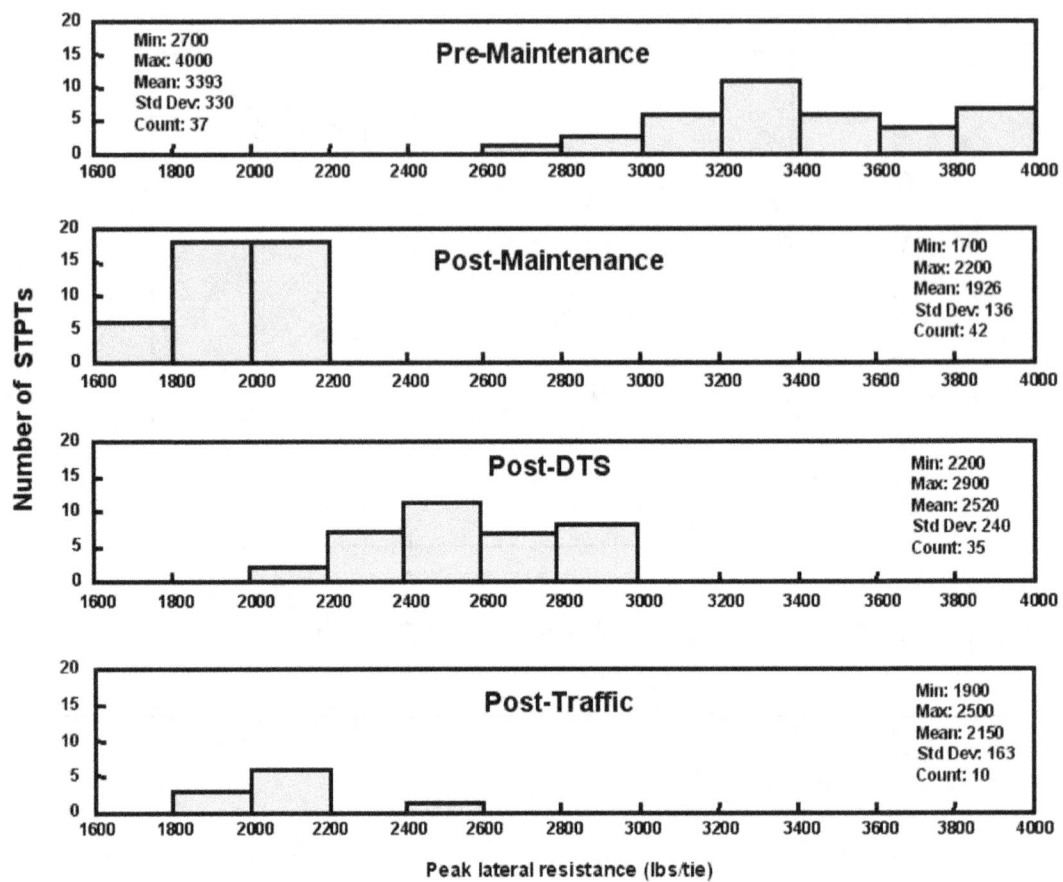

Figure B-4. STPT Concrete Tie Measurements on Amtrak

B.1.3 Torsional Stiffness Measurement

Torsional resistance is offered in the fasteners when the rail tends to rotate with respect to the ties i.e. resisting the in-plane bending, such as during buckling. Many rail fastener systems were tested, including two elastic fastener designs, as well as conventional wood tie plate/cut-spike systems. Laboratory and field tests were designed to evaluate pure rail rotation in fasteners while and to account for the tie rotation in the ballast (such as in the case of very torsionally rigid fasteners), respectively.

In the laboratory tests, the tie was held rigidly in a fixture, and equal and opposite loads were applied to the rail as shown schematically in Figure B-5 via hydraulic cylinders. The rail displacement was measured, and rotation was computed with the moment arm. All tests were conducted to at least a 5-degree rotation or until the onset of tie material crushing or failure. In cases of concrete tie fasteners, testing was also suspended if any insulators broke.

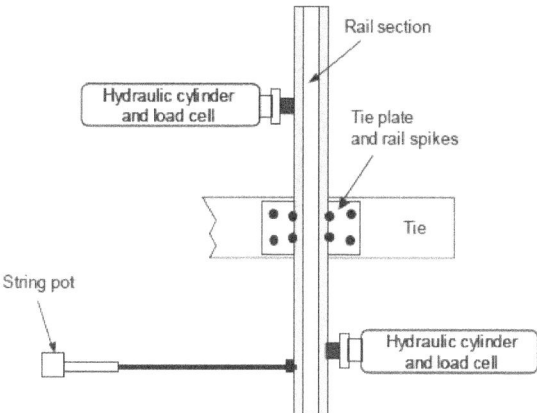

Figure B-5. Methodology for Torsional Stiffness Measurement

In the field tests, wood ties with cut-spike construction under existing service conditions were considered. The rails were cut into 40-inch sections. Two configurations of loading were initially devised, as shown in Figure B-6. Configuration A, was eliminated from further consideration because it applied a reactive lateral force on the fasteners. In Configuration B, the setup produces pure torque without lateral load on the fasteners. The test section chosen was 136 RE jointed rail with rail anchors on every other tie permitting an assessment of anchor influence.

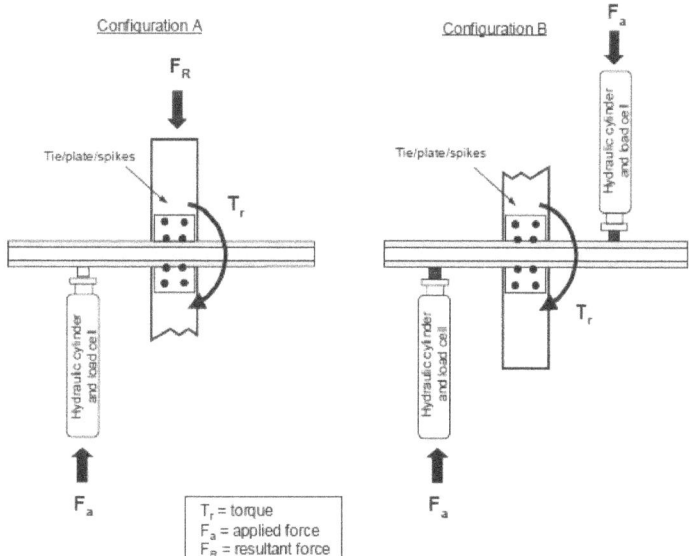

Figure B-6. Torsional Test Loading Configurations

B.1.4 Concrete Tie Test Results

Two elastic fastener systems were tested on concrete ties. Several test runs were performed to determine the effect of fully driven clips versus partially driven. No apparent difference in torsional resistance response was observed. Further comparisons were made between polyurethane pads versus rubber pads—again no significant differences were observed in the response curves. Overall, both systems exhibit a soft torsional resistance response in comparison with wood tie systems, as seen in the following section. The reason for the soft response is primarily due to the elasticity offered by the resilient pads and the softness of the insulators. This elastic attribute of the pads allowed the applied torque to be absorbed through insulator crushing rather than by the clips themselves, a fact observed through the insulators examination upon completion of each test run. The response for the two elastic fasteners was in the same range, suggesting fastening types on concrete ties play little role in determining the torsional resistance. Rather, the existence of insulators and pads appears to be the controlling factor. Typical results for the two systems are available in [9].

B.1.5 Wood Tie Results

Wood tie fastening systems tested produced a much stiffer resistance than the two elastic fattener systems on concrete ties. The specific fasteners tested on wood ties were a typical cut-spike construction, and elastic fastener with lock and screw spikes. Torsional response was compared for a variety of wood tie types and condition. Tie type and condition had little or no influence on the response characteristic, whereas the fastener type itself was a significant parameter. As expected, elastic fastener on wood ties were the stiffest, regardless of the use of lock versus screw spikes, followed by eight spikes per plate, four spikes per plate, and finally the softest wood tie characteristic was offered by two spikes per plate. Typical wood tie test results are available in [9].

B.2 Longitudinal Resistance

The longitudinal resistance of the track is defined as the resistance offered to the rail as it moves or tends to move in the longitudinal direction. In the case of curved track, this direction of movement can be considered to be tangential to the curve.

The resistance to the rail movement in the longitudinal direction is offered by the fasteners that hold the rail to the tie. The rail may move through the fasteners with the ties stationary, or the rail and the ties may move through the ballast when the fasteners provide a rigid connection. Figure B-7 shows the two scenarios schematically. To distinguish the two cases, one could define Figure B-7b to represent the fastener longitudinal resistance, as opposed to Figure B-7c, which represents the ballast longitudinal resistance. The net resistance to the rail can be a combination of these two resistances. That is, at small loads in the rail, initial rail movement can occur in the fasteners with respect to the tie. With increasing load, the rail may move along with the tie in the ballast. The behavior of the system is also dependent on the vehicle load.

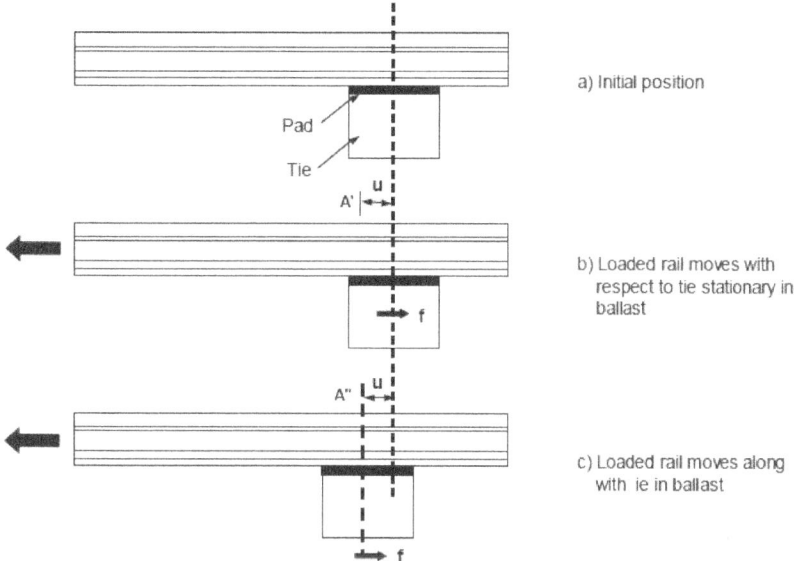

Figure B-7. Longitudinal Resistance Mechanism

The longitudinal resistance is an important contributor to the stability of CWR in the following manner:

- It controls the distribution of the rail longitudinal force and movements, thus affecting the rail neutral temperature.
- It provides the necessary reaction to train braking and acceleration forces.
- It controls the rail break gaps when pull-apart failures occur.

The factors that may significantly influence the longitudinal resistance include the following:

- Anchor effectiveness
- Toe load retention of fastener
- Pad shear stiffness
- Tie type and size
- Ballast type/condition
- Vertical vehicle load

The resistance can be expressed per fastener (per rail) or per tie (two fasteners per tie). The track longitudinal resistance is obtained by dividing the individual tie longitudinal resistance by the tie spacing.

$$\text{Track longitudinal resistance} = \frac{\text{Individual tie longitudinal resistance}}{\text{Tie spacing}}$$

The tie spacing in the above equation is the spacing of anchored ties (i.e., ties with fasteners attached to the rail). Wood ties with cut spike construction are generally anchored on every other tie. The spacing to be used in the foregoing equation should be twice the actual physical spacing of ties.

Figure 4-10 in Chapter 4 shows a typical response from a longitudinal resistance test. The resistance is idealized as linear-constant. The slope of the linear part is called longitudinal stiffness, k_f, whereas the constant f_o is called steady state value of the resistance.

B.2.1 Measurement of Longitudinal Resistance

The fastener longitudinal resistance can be measured in a laboratory using the pull-test, in which the rail is pulled through its fastener with the tie held rigidly by an external support. For the measurement of the ballast longitudinal resistance, a field setup is necessary. Figure B-8 shows a schematic of the arrangement used at TTC. A panel is cut and pulled longitudinally, reacted by adjacent portions of the track. The panel must have a minimum number of ties to prevent rotation over their long axes when subjected to loads via the rail.

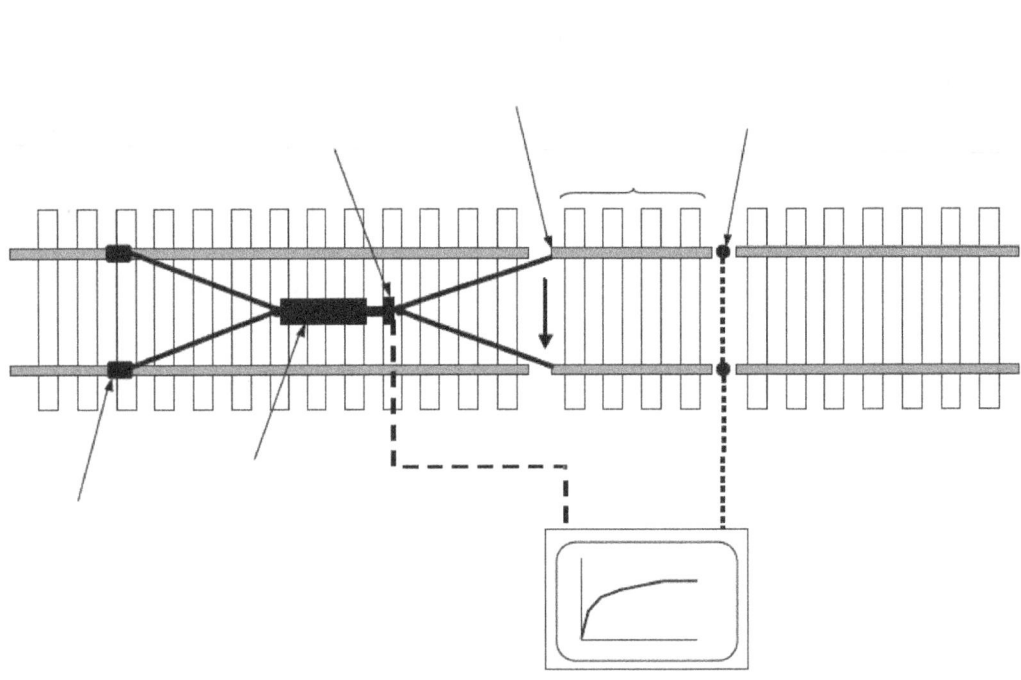

Figure B-8. Longitudinal Resistance Test Hardware

B.3 Tie-Ballast Friction Coefficient

The tie-ballast friction coefficient is determined using the STPT fixture for small vertical loads (≈ 1 kip) on the tie and the Track Loading Vehicle (TLV) for larger vertical loads (> 5 kips). The case with small vertical load on the tie is appropriate for calculating the loss of resistance in the central region between the trucks, which is the scenario for track buckling.

The friction coefficient is defined as the friction force at the bottom surface of tie interfacing the ballast divided by the weight of the tie. This interface is not like that between two bodies with plane surfaces exhibiting the typical Coulomb friction property. The ballast particles can physically penetrate the wood tie surface, thereby giving rise to an apparent friction coefficient larger than one in some cases where the tie has been in service over a long time. Under passing vehicles causing vibrations and vertical movement of ties, the friction coefficient falls off to a lower value compared to the value in a statically embedded state of the tie. This value of the friction coefficient during the tie movement is more appropriate for use in the dynamic buckling theory than the static value.

In the application of TLV to measure the friction coefficient, the TLV hydraulics applied vertical load on the tie and pushed the tie laterally (Figure B-9). Several ties with similar ballast shoulder and crib were tested with and without the vertical load. The difference between the peak values of the lateral resistance with and without vertical load on the tie divided by the vertical load on the tie was taken as the tie-ballast friction coefficient. The friction coefficient was also evaluated for a range of vertical loads as well. The friction coefficient at no applied vertical load on the tie, other than its self-weight, is also evaluated by testing a number of ties with and without ballast in shoulders and cribs. The difference in the peak values of the resistance divided by the tie weight was taken as the tie-ballast base friction coefficient. The values of the friction coefficient were found to be in the range of 0.9 to 1.2 for the wood tie and about 0.85 for the concrete tie, the ballast in both cases being granite.

Figure B-9. TLV Based Friction Coefficient Measurement

B.4 Rail Force and Neutral Temperature

A critical review of the techniques to measure rail force and neutral temperature revealed that only two techniques are available to measure rail force with acceptable accuracy and reliability. These are the strain gage and the Rail Uplift Technique (RUD). The strain gage is well known in the literature and has been used extensively in a number of measurements for the FRA/Volpe projects on CWR. The authors [14] originally conceived the Rail Uplift Technique and extensive tests were carried out at TTC to validate the technique.

Strain gage technique needs rail cutting to provide a "reference" for the absolute force. The RUD can yield absolute force after calibration which is not site dependent. The device has an analytic basis as explained in the following paragraph. The disadvantage of the device is that is requires freeing the rail from anchors or fasteners, which makes it somewhat cumbersome in application.

B.5 Rail Uplift Device

Vertical deflection depends on the magnitude of the rail longitudinal force when rail-ties fasteners are removed over some length, restrained vertically at the ends of the freed portion, and subjected to a concentrated uplift load at the center. Clearly, longitudinal compressive load will increase the deflection of the beam-column, and tensile force will reduce it. For a given length of rail, the vertical force required to produce a specified deflection is a measure of these rail forces. The concept implementation is based on the fact that the rail can be held pinned at the two end points by the wheels of a rail car. This automatically fixes the length of the rail and boundary conditions at the ends of the rail beam. The spikes and anchors between the inner wheels of the two trucks of the car must be removed. Figure B-10 shows schematically the rail uplift method.

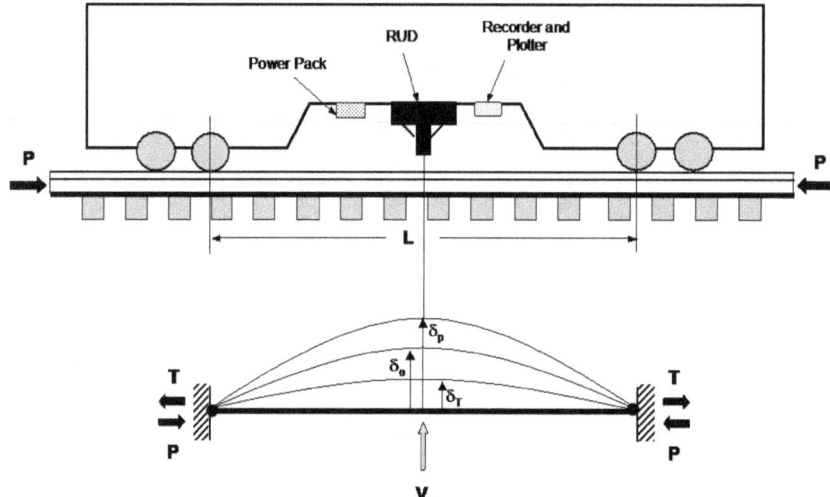

Figure B-10. Rail Uplift Device Principle

An analytic model was developed to relate the applied vertical force and deflection to longitudinal rail force. The model showed that the relationship between the applied vertical force and the rail longitudinal force is linear. A device can be operated without rail buckling for small deflections (< 2") over the practical range of the rail longitudinal force ±120 kips and the rail temperature range 40 °F to 100 °F. Instead of correlating the deflection with the rail force for a constant applied vertical load, one can correlate the required uplift force for a fixed vertical deflection with the rail longitudinal force.

B.6 Validations

Tests were conducted at TTC, Pueblo, CO, on a tangent and a 5-degree curved track to validate the technique. A special instrumentation car with inner wheel spacing of 340 in was adapted to provide a maximum vertical force of 30 kips. The test sections were instrumented with strain gages to measure the rail force. The variation in the rail force was achieved through destressing at reasonable high neutral temperatures for tensile loads and by means of artificial rail heating for compressive force levels. Figure B-11 shows the mean regression line for all the test data and the theoretical prediction, which are in good agreement. Test validations were also performed on a 5-degree curve.

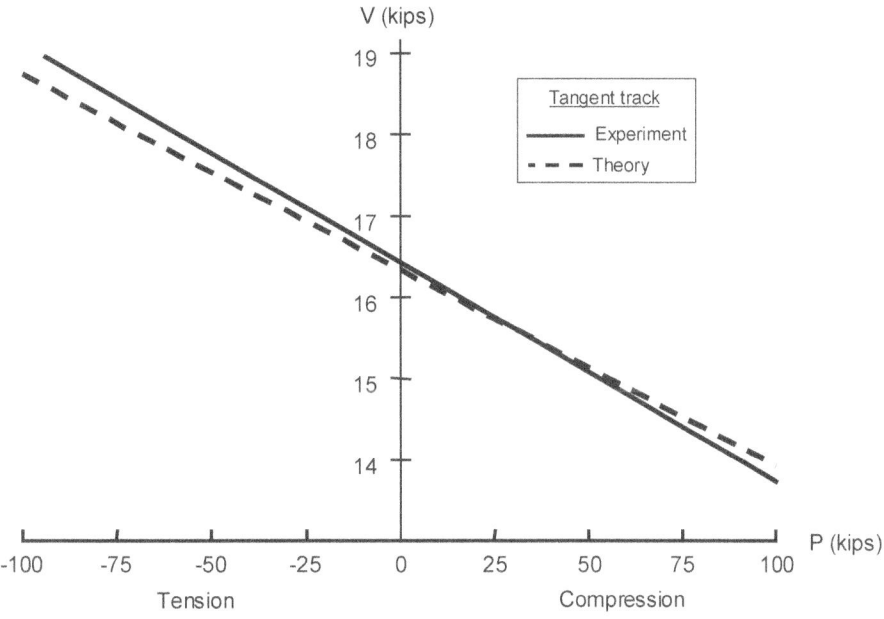

Figure B-11. Rail Uplift Principle Validation

B.7 Field Application

The RUD has been applied in the field to map the neutral temperature on BNSF railroads [15]. This was performed using AAR's TLV, which was modified to apply the uplift force and measure the uplift deflection (Figure B-12). Several thousand feet of the track was mapped, including a tangent and 5-degree to 6-degree curved section. Figure B-13 shows an example of the measured rail neutral temperature variation.

Figure B-12. TLV Based Rail Uplift Measurements

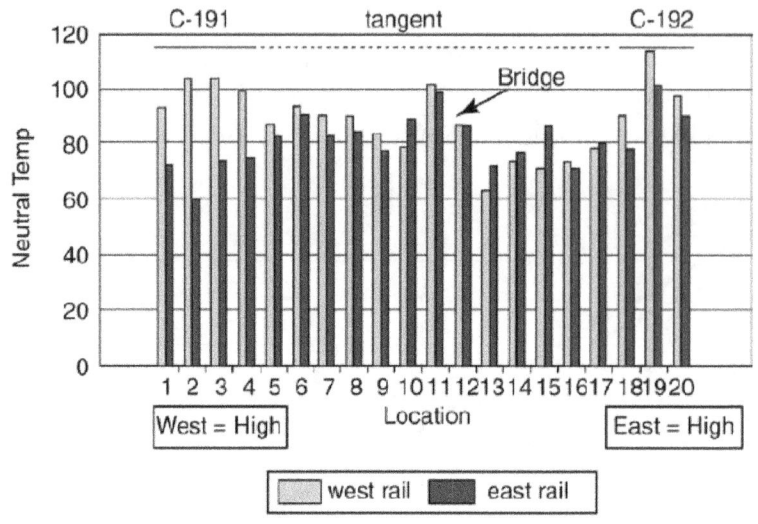

Neutral Temperature for Section 2

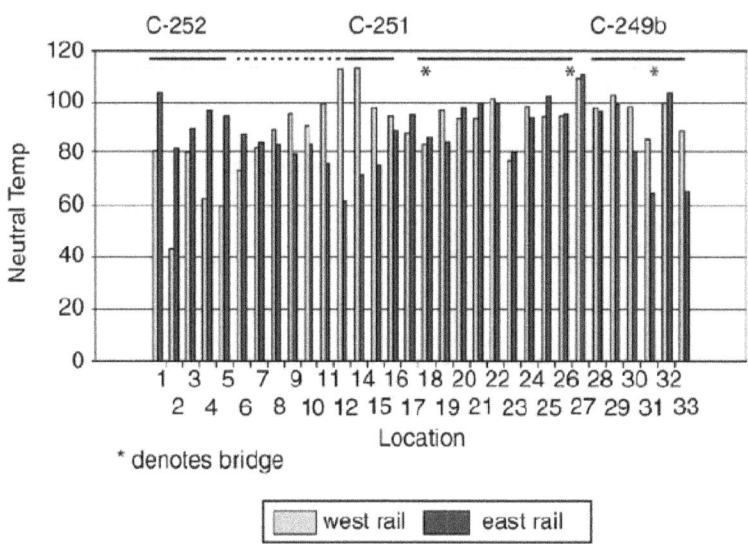

Neutral Temperature Variation for Section 3

Figure B-13. Examples of Neutral Temperature Measurements Using the Rail Uplift Device (Locations Spaced 300 to 500 ft Apart)

Appendix C.
Theoretical Basis of CWR-Risk

This appendix will present analytic details of evaluating the probability of buckling on the basis of the convolution integral used in Chapter 3. Figure C-1 shows the distributions of load and strength. These are represented by ψ and ϕ respectively. If the strength curve (ϕ) is totally separated to the right of the load curve, ψ, no interference will exist between the curves. This means that the strength is always greater than the load, hence be no buckling will occurred. In cases where there is some interference, as seen in Figure C-1, a range of temperatures (A to D) exists in which load can exceed strength, thereby giving rise to buckling situations. To evaluate the buckling probability at a temperature in the interference range, the convolution integral is used. To facilitate an understanding of this integral and its evaluation, it is convenient to represent the load by x and the strength by y.

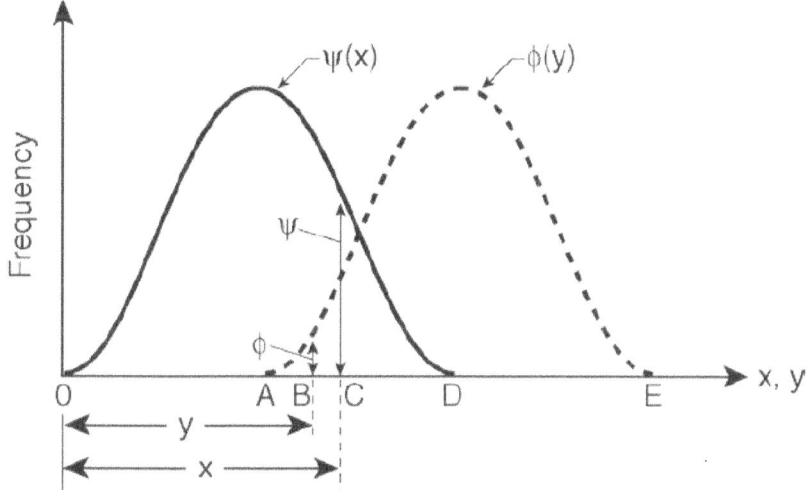

Figure C-1. Frequency Distributions of Load and Strength, and the Interference Zone

$x = $ load ($T-T_N$),

where

T = rail temperature,
T_N = neutral temperature

$\psi(x)$ = load frequency (see Figure C-2), represents the probability of x being equal to a specified value X by

$P(x = X) = \psi(X)$

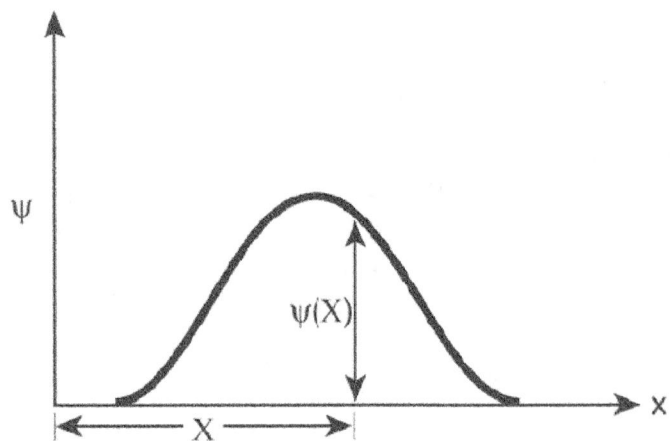

Figure C-2. Frequency Distribution Function of Load

The probability load function is defined as

$$\Psi(x) = \int_0^x \psi(x)\,dx \quad \text{(see Figure C-3)}$$

and represents the probability of x being equal to or less than a specified value X by

$$P(x \le X) = \Psi(X)$$

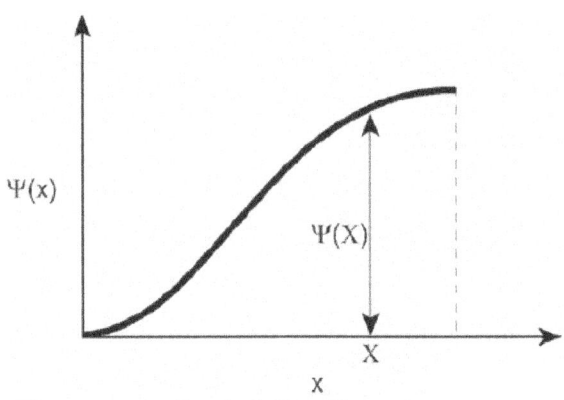

Figure C-3. Probability Function of Load

y = strength, ΔT_{all}

$\phi(y)$ = strength frequency (see Figure C-4)

and represents the probability of y being equal to a specified value Y by

$$P(y = Y) = \phi(Y)$$

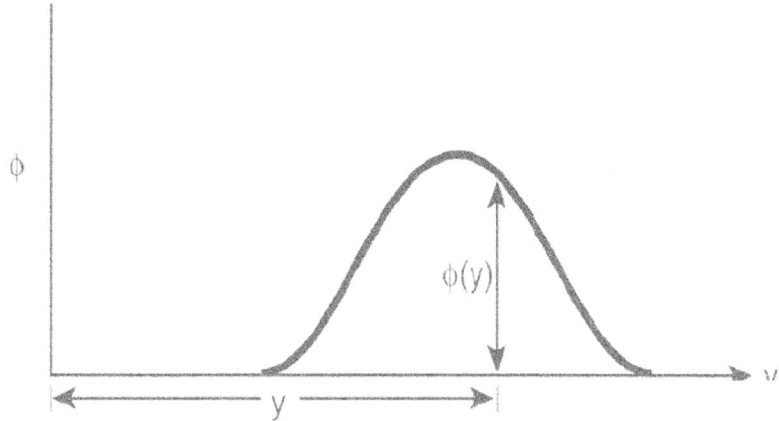

Figure C-4. Frequency Distribution Function of Strength

The probability strength function is defined as

$$\Phi(y) = \int_0^y \phi(y)\, dy \quad (\text{see Figure C-5})$$

and represents the probability of y being equal to or less than a specified value Y by

$$P(y \le Y) = \Phi(Y)$$

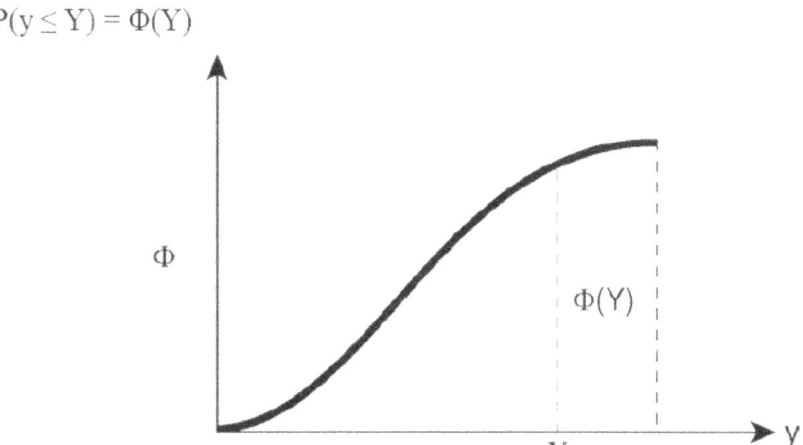

Figure C-5. Probability Function of Strength

Referring to Figure C-1, consider point C representing the load x and point B representing the strength y, which is less than x. The probability of x ≥ y is, by the product law, $\phi(y) * \psi(x)$. Now let C vary from B to D. The total probability of C assuming greater or equal values of B, between B and D, equals

$$\int_{B}^{D} \psi(x)\, \phi(y)\, dx$$

The above integral y and $\phi(Y)$ are fixed; only x and hence $\psi(x)$ are varied.

Hence in the interval BD, the $P(C \geq B) = \int_{B}^{D} \psi(x)\, \phi(y)\, dx$

To obtain the sum total probability of load exceeding strength, let B vary from A to D

$$P\,(C \geq B) \text{ for interval } AD = \int_{A}^{D} \left\{ \int_{B}^{D} \psi(x)\, \phi(y)\, dx \right\} dy$$

$$= \int_{A}^{D} \left\{ \int_{0}^{D} \psi(x)\, \phi(y)\, dx - \int_{0}^{y} \psi(x)\, \phi(y)\, dx \right\} dy$$

$$= \int_{A}^{D} \left\{ \phi(y) - \phi(y) \int_{0}^{y} \psi(x)\, dx \right\} dy$$

Since $\quad \int_{0}^{D} \psi(x)\, dx = \left[\Psi(x) \right]_{0}^{1} \;=\; 1$

Also $\quad \int_{0}^{y} \psi(x)\, dx = \left[\Psi(x) \right]_{0}^{y} \;=\; \Psi(y)$

$$\boxed{\; \text{P(Buckling)} \;=\; \int_{A}^{D} \phi(y) \left[1 - \Psi(y) \right] dy \;}$$

The integral involves only one variable, y, and is confined to the interference zone.

Numerical Work

For numerical work, points A and D (A being the starting point of the strength curve and D being the end point of the load curve which define the range of temperature in the interference zone) are identified in the CWR-SAFE program. The interval A to D is divided into an even number of intervals (n). Define the intervals between A and D: y(1), y(2), ... y(n+1).

The program identifies the non-zero values shown as √ in the following matrix. The last value ψ and the first value of φ are zero as they represent points A and D respectively.

$\psi(1) = \sqrt{}$ $\phi(1) = 0$
$\psi(2) = \sqrt{}$ $\phi(2) = \sqrt{}$
....
$\psi(n+1) = 0$ $\phi(n+1) = \sqrt{}$

The program computes

$\Psi(1) = 0$
$\Psi(2) = [\ \psi(1) + \psi(2)\ ] / 2$
$\Psi(3) = \{\ \Psi(2) + [\ \psi(2) + \psi(3)\ ] / 2\ \}$
$\Psi(n+1) = \{\ \Psi(n) + [\ \psi(n) + \psi(n+1)\ ] / 2\ \}$

The program defines $Z(i) = \phi(i)\ [\ 1 - \psi(i)\ ]$ for $i = 1, 2, \ldots n+1$

which is the integrand in P(Buckling) shown above.

The program computes by a trapezoidal rule $\int_{1}^{n+1} Z(i)dy$

i.e., $\int_{1}^{n+1} Z(i)dy = \dfrac{Z_1}{2} + \sum_{2}^{n} Z(i) + \dfrac{Z_{n+1}}{2}$

Appendix D.
Measurement Requirements and Sample Rates for Lateral Resistance and Neutral Temperature

Appendix B describes the methods and hardware to measure the two key parameters of lateral resistance and neutral temperature. The measurements are needed to provide input to CWR-SAFE for monitoring and controlling the parameters for safety assurance in revenue lines. The measurements are typically required for two scenarios. One is over a defined small segment of track for use in the deterministic approach for buckling safety evaluation. The second one is for the determination of distribution of the parameter over a territory for use in the probabilistic approach.

D.1 Lateral Resistance

D.1.1 Sampling Rate Basis

A number of lateral resistance measurements were made under the FRA research program on different track territories, including those of Amtrak, Union Pacific and Transportation Technology Center using the Single Tie Push Test (STPT) device. In many of these measurements, a 50-foot long test zone was considered as a characteristic length (approximately the truck center spacing for vehicles where buckles typically occur). A number of test samples were taken over each of the 50-foot sections, and a mean value of resistance was identified for each test segment. For a wood tie track with about 20" tie spacing, there are 15 testable ties since only alternate ties can be tested. For the concrete tie with 24" spacing, the test fixture tends to react on two adjacent ties on each side of the test tie, thereby limiting the maximum number of testable ties to about 9 per a 50-foot long test zone. The mean value of 9 ties in the 50-foot test zone is considered as a representative number for the resistance in the zone.

Whereas for research purposes, one can test 9 ties in concrete and 15 ties in wood in a 50 foot test zone; this is not considered to be practical for rapid safety evaluations by the industry. It is also necessary to minimize excessive track disturbance to the track. Therefore, the problem of minimum required number of tie samples in the zone should be addressed. The mean, F_n, of the selected number of samples, n, must be reasonably close to the average resistance, F_0, obtained by testing the maximum possible number of ties in the zone. A tolerance for the difference $e = F_0 - F_n$ must be established first.

Tolerance

Table D-1 gives a sample of allowable temperature increase for lateral resistance values in the range of 1500 to 3000 lb/tie. This table also shows the allowable temperature values for a resistance drop of 15 percent from the nominal values in the range. Table D-2 shows other parameters assumed in this analysis.

Table D-1. Sensitivity of Allowable Rail Temperature with Respect to Lateral Resistance and Peak Value

Nominal Lateral Resistance (lb)	Tangent Track			5-Degree Curve		
	ΔT_1	ΔT_2	δT	ΔT_1	ΔT_2	δT
1500	94.63	92.21	2.42	82.56	78.98	3.58
2000	99.49	96.64	2.85	88.74	85.3	3.44
2500	104.01	100.64	3.37	93.98	90.1	3.88
3000	108.42	104.45	3.97	98.86	94.48	4.38

ΔT_1 = Allowable Temperature for Nominal Value of Resistance (°F)
ΔT_2 = Allowable Temperature for 85% of Nominal Resistance (°F)
δT = Change in Allowable Temperature (°F)

Table D-2. Assumed Parameters

Data Inputs	
Rail Size:	Area 136
Tie Type:	Wood
Tie Weight (lb):	200
Tie Spacing (in):	20
Ballast Type:	Granite
Tie-Ballast Friction Coefficient	1.2
Torsional Resistance (kips/in./rad):	100
Longitudinal Stiffness (lb/in./in.):	200
Foundation Modulus (psi):	6000
Peak Resistance (lb/in.):	100
Misalignment (in.):	0.625
Misalignment Half-Wavelength (in.):	180.6
Rail Neutral Temperature (°F):	80
Maximum Rail Temperature (°F):	140
Vehicle Type:	Hopper

It is seen that the corresponding change in the allowable temperature does not exceed 5 °F when the change in lateral resistance is under 15 percent. It will be assumed that the buckling temperature error of ±5 °F is tolerable in practical applications. The change in allowable temperature tends to reach or exceed 5 °F only at high resistance values (>3000 lb/tie), but such high resistance situations generally do not create buckling problems. Therefore, it will be concluded that a tolerance of ±15 percent in the lateral resistance measurement is permissible for the wood tie track. The same conclusion can also be reached for the concrete tie track.

D.1.1.1 Sampling Rate

The required sampling rate for a permissible tolerance of ±15 percent resistance will be determined using the statistical equations of confidence levels and intervals. As stated, a large number of resistance measurements were made, and it was found that these can be reasonably represented by normal distributions, as shown in [10]. A normal distribution is represented by the two parameters, namely the mean and the standard deviation.

• Mean value, F_0

$$F_0 = \sum_{i=1}^{N} F_i / N \tag{D-1}$$

F_i represents the peak resistance value of the ith tie, and N is the number of ties used in the 50 ft zone, which is at least 9.

• Standard deviation, σ

$$\sigma = \left\{ \sum_{i=1}^{N} (F_i - F_0)^2 \right\}^{1/2} \bigg/ (N-1) \tag{D-2}$$

The determination of the sampling rate is as follows. Suppose that instead of testing N = 9 or greater number of ties in the 50 foot zone to determine the average resistance, F_0, assigned to the zone, one opts to test a much smaller number of ties, n << N. The mean of the n number of test tie data, F_n, will be different from F_0. For a maximum permissible difference of ±e ($e = |F_0 - F_n|$) or for a

$$\text{Confidence Interval} = (F_n - e \leq F_0 \leq F_n + e), \tag{D-3}$$

one can establish a relationship between the number of samples and the probability of finding the mean value within the confidence interval. The sampling rate, n, is given by the well-known formula [24] in statistics

$$n = (C\sigma/e)^2 \tag{D-4}$$

Here C is a constant which depends on the required Confidence Level. If the Confidence Level is designated by Y, the relationship between Y and C is shown in Table D-3. This table is valid for normal distributions and has been well documented in [24].

It is usual practice in sample sizing to examine the three Confidence Levels presented in Table D-3. The higher the stipulated Confidence Level, the larger the value of C and the sampling rate increases with the square of C. A 90 percent Confidence Level is adequate in many of the engineering applications. This will be recommended for CWR track safety monitoring. In view of the built-in margin of safety in the proposed safety limits, a higher Confidence Level may not be justified as it increases the monitoring requirements for the lateral resistance.

Table D-3. Relationship between Confidence Level and Parameter C (for Normal Distribution)

Y (Probability or Confidence Level)	C (Parameter)
90%	1.645
95%	1.960
99%	2.576

For the assumed tolerance e = 15 percent and 90 percent confidence level, the sampling rate n is calculated from equation D-4 for a number sites with historical data collected using the traditional method of testing 9 or more ties in 50 foot zones.

Table D-4 provides the required sampling rate for tracks tested at TTC, UP, and Amtrak. Wood and concrete ties and slag and granite ballast materials were considered. Tamped, stabilized under revenue traffic, and artificially stabilized conditions were also variables in the data.

Table D-4. Examples of Sampling Rate Calculations

Site Location	Track Type	Zone #	Initial Condition	MGT	Measured Mean (lb)[1]	Std Dev	Required Sampling Rate[2]
TTC Ref [10]	Wood	1	Slag Ballast	.1	1469	258	4
		2	Granite Ballast, Trafficked	25	1993	397	5
		3	Slag Ballast, Trafficked	25	2374	357	3
		4	Slag Ballast	0	1038	148	2
		5	Slag Trafficked	100	3176	560	4
		6	Granite Trafficked	25	2206	411	4
UP Ref [25]	Wood	1	Granite Tamped	0	3300	460	2
		2	Granite Trafficked	.4	3371	420	2
		3	Granite Trafficked	.6	3430	510	3
Amtrak Ref [22]	Concrete	1-a	Granite Consolidated	-	3483	275	1
		1-b	Granite Tamped	0	2000	128	1
		1-c	Granite Stabilized	0	2611	231	1
		2-a	Granite Consolidated	-	3030	197	1
		2-b	Granite Tamped	0	1900	145	1
		2-c	Granite Stabilized	0	2355	217	1
		3-a	Granite Consolidated	-	3610	239	1
		3-b	Granite Tamped	0	1885	100	1
		3-c	Granite Stabilized	0	2575	198	1
		4	Granite Tamped	0	1905	152	1

[1] *Measured Mean was for 9 or more ties in the case of 50 foot wood tie sections and 5 for 50 foot concrete tie sections.*
[2] *Required Sampling Rate is for an error within 15 percent of measured mean with 90 percent confidence value.*

Although some variability occurs in the results for the sampling rate, it appears that a sampling rate of one tie for a 50 foot concrete section and 2 ties for a 50 foot wood tie section can be recommended to determine the resistance with an error of 15 percent with a Confidence Level of 90 percent.

D.1.2 Evaluation of Lateral Resistance Statistical Distribution on a Territory

To determine the distribution of lateral resistance that is applicable over a large territory of a railroad division, it is not necessary to measure the resistance on every 50 foot track segment. Adequate data can be collected by first identifying weak locations for selective measurements and super imposing data from the remaining normal track conditions/locations.

The representative weak locations include the following:

    A. New or recently maintained segments
    B. Stabilized segments before full speed restoration
    C. Degraded track (poor ballast and tie and settling subgrade conditions)
    D. High curvature CWR track with excessive radial movement
    E. Locations with previous buckling incidents

A and B type track segment resistances can be estimated using CWR-INDY's lateral resistance estimator (see Equation 3.5 in Chapter 3). This is because, for new and recently maintained tracks, the lateral resistance is fairly uniform, and CWR-INDY's predictor is fairly good for such conditions. Conversely, where larger variations are expected, such as in C, D, and E type segments, the STPT method will be required. The sampling rates for those measurements shall be as per the recommendations made in Subsection D.1.1.

To generate data for normal or typical track segments, isolate representative sections of the following four categories: wood and concrete, each with curvatures from 0° to 4° and from 5° to 9°. A representative section is one that exhibits similar characteristics as the majority of the territory (i.e., one which has similar ballast, tie type, ballast section, and MGT). The length of the representative section can be as large as a quarter mile. Using the recommended sampling rates (i.e., one tie/50 foot of concrete tie and two ties/50 foot of wood tie, the lateral resistance measurement should be carried out).

Combining the measurements for the weak locations with those for the normal section, one can construct a database for the territory and deduce the statistical distribution (Figure D-1). The database can be updated at selected locations on an annual basis, particularly for the weak locations during early summer conditions. For normal segments, the measurement of resistance must be made only once.

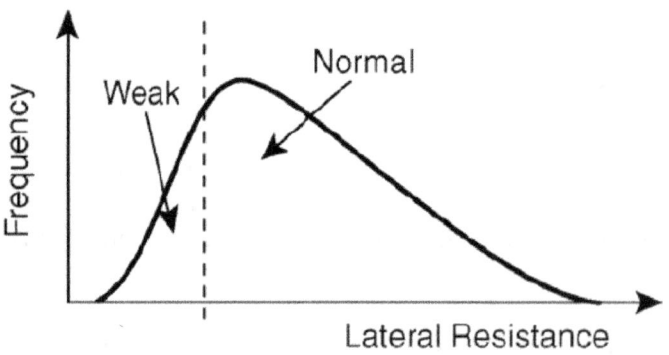

Figure D-1. Statistical Distribution of Resistance

D.2 Neutral Temperature

D.2.1 Sampling Rate Basis

Several tests were carried out on the evaluation of neutral temperature on revenue lines and TTC tracks under the sponsorship of the FRA. The tests were performed for different purposes, including the following:

1. Mapping of neutral temperature on long stretches of track (5° and 6° curves and tangent on BNSF, [15])
2. Determination of daily and seasonal neutral temperature changes in revenue lines and at TTC
3. Validation of longitudinal restraint model for winter rail break
4. Restressing CWR at the required neutral temperature (on TTC track, [16])

The tests for item 1 revealed that over several thousand feet of track at BNSF test sites the neutral temperature varied in the range of 110 °F to 60 °F. The measurement spacing varied from 300 to 500 ft. Hence, it is inferred that a 60 °F minimum is achievable on revenue lines. This number is used in the safety implementation presented in Chapter 6.

The tests for item 2 revealed that daily temperature fluctuation caused by rail longitudinal movement through fasteners and anchors could result in a neutral temperature change of several degrees (< 10 °F). These tests also revealed the influence of curve breathing on neutral temperature.

The tests carried out at TTC with the objective stated in item 3 showed that the longitudinal resistance in a typical wood tie track varies in the range of 15 to 30 lb/in, depending on the anchoring pattern and other factors.

To estimate the potential variations in the neutral temperature along a rail, the following analysis has been used. Consider the equilibrium of a single rail of length ℓ (Figure D-2).

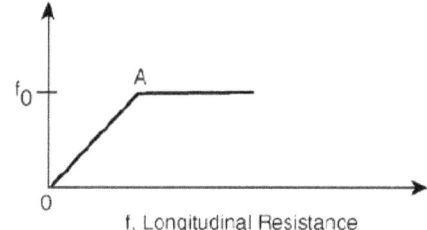

a) Longitudinal Resistance Characteristic

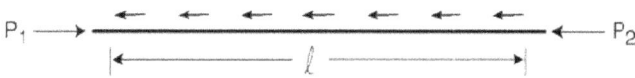

b) Equilibrium of a Single Rail

Figure D-2. Evaluation of Change in Rail Longitudinal Force

From equilibrium considerations

$$f_0\ell = P_1 - P_2 \tag{D-5}$$

where f_0 is the maximum longitudinal resistance (per rail per unit length). By definition of rail neutral temperature at locations 1 and 2.

$$P_1 = AE\alpha\ (T_R - T_{N1}) \tag{D-6}$$

$$P_2 = AE\alpha\ (T_R - T_{N2}) \tag{D-7}$$

where

A = the cross sectional area of a single rail,
E = Young's Modulus,
α = the coefficient of thermal expansion, and
T_R = rail temperature.
$T_{N1,2}$ = rail neutral temperature at locations 1 and 2.

From equations D-5, D-6, and D-7, it is seen that

$$(T_{N1} - T_{N2}) = \Delta T_N = f_0\ell/AE\alpha. \tag{D-8}$$

Here ΔT_N = change in neutral temperature between locations 1 and 2.

This equation represents the possible variation of neutral temperature between two locations on one rail, which are a distance, ℓ, apart. If an error of 5 °F is permitted in the neutral temperature, then the maximum permissible distance between two measurement points is

$$\ell_{max} = (5AE\alpha/f_0)$$

The error of 5 °F in neutral temperature is the same value used for the error in the allowable temperature, T_{all}, due to the 15 percent permissible error in the lateral resistance. The longitudinal resistance, f_0, is typically 25 lb/in. For typical rail cross-

D-7

sections, it can be shown that $\ell_{max} \approx 50$ ft. In practice, this will be larger than 50 ft since the longitudinal resistance is determined by a point below A on the initial linear line, particularly if the rail is uncut (Figure D-2).

If the neutral temperature measurement point is chosen at P_1, then the permissible ℓ_{max} on the left side of location 1 in Figure D-2 will be the same as 50 ft. It can therefore be recommended that for every 100 ft rail section, one measurement point be used for the stated maximum permissible error of 5 °F. This is further confirmed from the data analysis carried out in the following paragraph.

Table D-5 shows an example analysis of data collected on TTC track during the winter rail break tests. The sample number of strain gage locations used in the tests to measure the neutral temperature was 8 for every 100 foot. The mean of the 8 strain gage values is considered as the value against which the required sample rate for a maximum tolerable error of 5 °F is evaluated. It can be recommended that one sample over a 100 foot rail section is adequate to define the neutral temperature with 90 percent confidence limit for the maximum tolerable error of 5 °F.

Table D-5. Required Sampling Rate for 100 ft Long Tangent Track (90% Confidence Limit, 5 °F Error on the Mean Value).

Test	Anchoring	T_N^1 Measured Mean (°F)	Std Dev (°F)	Required Sampling Rate
WRB1	EOTA	81.2	2.5	1
WRB2	EOTA	78.8	2.1	1
WRB3	EOTA	102.2	2.6	1
WRB4	EOTA	118.7	2.4	1
WRB5a	ETA	98.7	0.8	1
[1] Number of data points = 8 ETA = Every Tie Anchored EOTA = Every Other Tie Anchored WRB = Winter Rail Break Test				

D.2.2 Evaluation of Neutral Temperature Statistical Distribution on a Territory

To determine the distribution of the neutral temperature on a territory, the following potential weak spots should be identified:

A. High degree curves exhibiting large radial (spring time movement)
B. Rail repaired/neutral temperature readjusted locations
C. Recently maintained locations (lined or lifted)
D. Bottom of steep grades, and locations with high traction and braking
E. Locations experiencing rail movement through the fasteners
F. Locations with a buckling history
G. Locations with rail fracture/pull apart history

D-8

For the weak locations (A to G), a one time monitoring with strain gages or a rail uplift device (RDU) will be required at the recommended minimum sampling rate. The monitoring is aimed to determine the lowest T_N value and to assure 60 °F minimum neutral temperature. Neutral temperature data should also be gathered on normal or typical track segments representative of a territory. This can be done on a quarter mile tangent wood and concrete segment and a quarter mile of 5° wood and concrete tie segment, which are good representations of the territory in terms of CWR parameters, such as age of rail, fastener type and condition. Perform measurements on these segments on both rails either via strain gauging or via RUD in accordance to the sampling methodology of Section 2.1. Thus, the statistical distribution on CWR neutral temperature can be assembled by combining the A-G data and the normal data for probabilistic evaluations.

Appendix E.
Rail Destress Force Indicator (RDFI)

Rail destressing and restressing are routine operations for the railroads, required whenever rail is intentionally cut in summer or fractured due to fatigue cracks and excessive longitudinal tensile forces developed in winter. Intentional cuts in summer will relieve excessive compressive forces and avoid potential buckles under high longitudinal compressive loads. Rail restressing is required to install or reweld the rails at a desired neutral temperature.

To install and reweld the rail at the desired neutral temperature, railroads depend on rail heating or hydraulic tensors. Rail heating can be artificial using propane gas, although this practice is being discontinued because of safety reasons. Solar heating is resorted to in many cases.

One specific aspect of destressing is de-anchoring or unfastening the rail from the ties after cutting the rail. This will ensure that the rail compressive force is eliminated, not only at the cut location, but also through a substantial portion of the rail on either side. Some of the railroad practices recommend a de-anchoring length from 5 rail lengths or 195 ft to longer lengths on either side of the cut location, but the guidelines are vague and not rationally formulated. After de-anchoring over whatever length the track crew decides on, a gap adjustment is performed at the prevailing rail temperature that will yield the desired neutral temperature under tensor application. The gap adjustment includes a 1" welding allowance.

Figure E-1 shows the definition of rail gap and the sequence of events in summer rail cutting and rewelding at the desired neutral temperature. Figure E-2 shows a similar situation when winter rail break occurs, and tensors are applied to restress the rail to the desired neutral temperature.

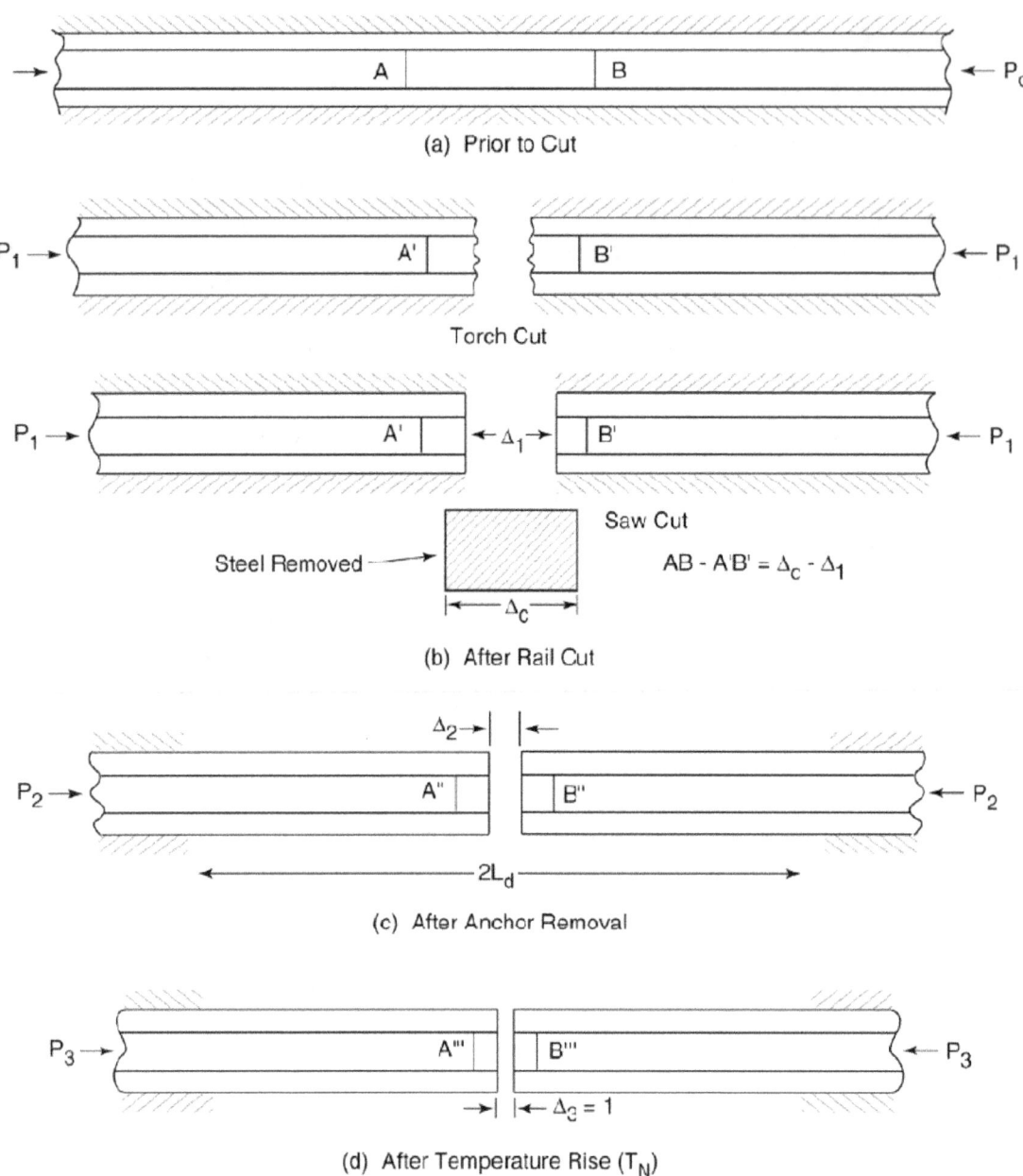

(a) Prior to Cut

Torch Cut

Saw Cut

Steel Removed →

$AB - A'B' = \Delta_c - \Delta_1$

(b) After Rail Cut

$2L_d$

(c) After Anchor Removal

$\Delta_c = 1$

(d) After Temperature Rise (T_N)

**Figure E-1. Definition and Measurement of Cutout Length and Rail
Gap in a Summer Rail Cut Scenario**

In the summer scenario, an optimum amount of steel must be removed and the rail de-
anchored over a required length so that a gap of about 1″ is achieved at the desired rail
welding (neutral) temperature. Likewise, in winter after rail fracture or cuts, the correct
amount of steel must be added, and/or an optimum tensor force must be applied to reduce
the gap to about 1″ for rewelding after the anchor removal over an optimum length.

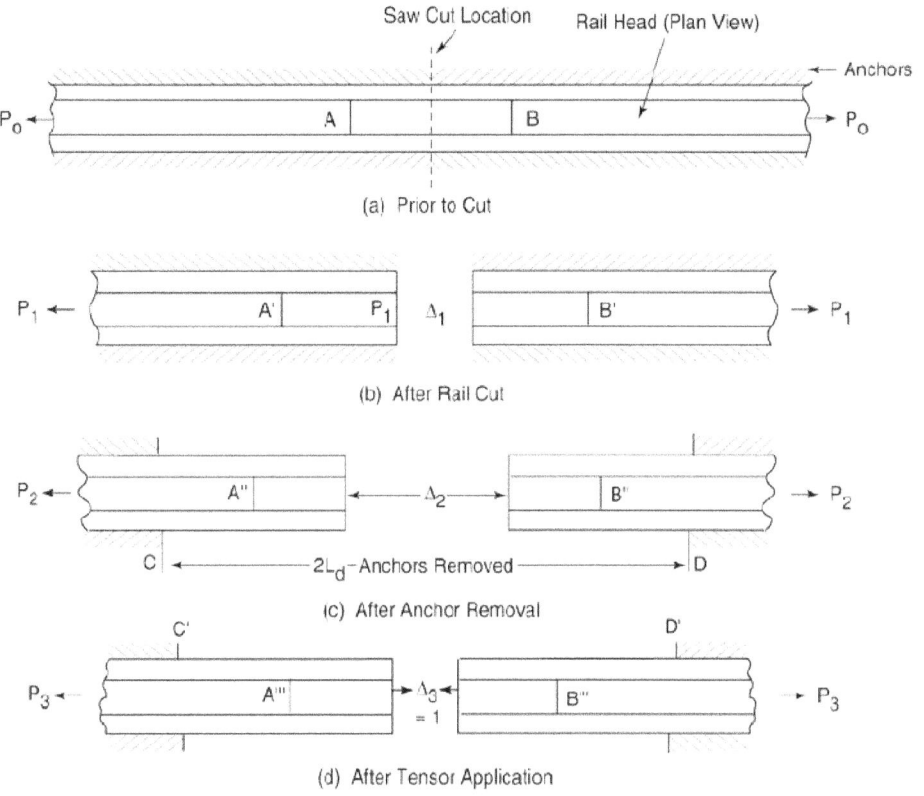

Figure E-2. Definition and Measurement of Rail Gap (Tensor Application)

A need to develop an instrument to facilitate the destressing and restressing operations in an optimal manner exists. Large errors can occur in the resulting neutral temperature if these operations are not performed properly. (For example, see Section 6.3.2). A breadboard device has been previously developed and tested, which consists of a strain gage and a temperature probe and uses a calculation program to output the optimum de-anchoring or unfastening length and the gap for a prescribed target neutral temperature. Initial rail movement due to cutting is a manual measurement which is also required as input into the bread-board device.

This measurement is made using two scribe marks on the rail head on either side of the cut and can be automated for easy use by the crew in future devices. The computer program back-calculates the longitudinal resistance at the site using the rail force from the strain gage and the resulting rail displacement from the rail cutting. On the basis of the analytic model presented in [16], the device outputs the required de-anchored length and the final adjusted gap length for a desired neutral temperature which is also an input entry to the device. The strain gage can be used to monitor the correctness of the newly set neutral temperature, as well as for subsequent monitoring to track how neutral temperature changes. Current research is addressing the development of such a prototype device in a field-functional version known as Rail Destress Force Indicator.

List of Abbreviations and Acronyms

AAR	Association of American Railroads
CWR	continuous welded rail
DTE	Equivalent MGT in Dynamic Track Stabilization
DTS	Dynamic Track Stabilization
EOTA	every other tie anchored
ERRI	European Rail Research Institute
ETA	every tie anchored
FRA	Federal Railroad Administration
MGT	million gross tons (traffic)
NAL	net axle lateral load
RDFI	Rail Destress Force Indicator
RUD	Rail Uplift Device
SF	safety factor
STPT	Single Tie Push Test
TCS	truck center spacing
TTC	Transportation Technology Center
UIC	Union Internationale des Chemins de Fer (Union of International Railways)

List of Symbols

Ω	Energy required to buckle the track
ψ	Load frequency
β	Constant values for adjust the lateral resistance
α	Rail steel coefficient of thermal expansion
θ	Rotation angle
σ	Standard deviation
ϕ	Strength frequency
μ_f	Tie-ballast friction coefficient
δ_o	Initial misalignment amplitude
τ_o	Torsional stiffness per rail fasteners
τ	Applied torque per fastener
ΔT_{Bmax}	Upper buckling temperature increase
ΔT_{Bmin}	Lower buckling temperature increase
ΔT_r	Rail temperature increase
$2L$	Buckling wavelength
$2L_o$	Misalignment wavelength
A	Rail cross sectional area
C_d	the ratio of existing crib depth to the full depth
C	Confidence level constant
E	Rail steel modulus
e	Tolerance (permissible error)
e_b	Expected number of buckles per year
e_R	Strain gage reading at the rail temperature
f	Longitudinal resistance
F_b	Resistance from tie bottom
F_e	Resistance from shoulder
F_L	Limiting lateral resistance
F_n	Mean value of n samples
F_P	Peak lateral resistance
F_s	Resistance from tie sides
F_{P0}	Tamped resistance
ΔFP	Increment due to consolidation
I_{yy}	Rail section moment of inertia about the lateral axis through centroid
I_{zz}	Rail section moment of inertia about the vertical axis through centroid
k_f	Longitudinal stiffness
kip	Unit of force equaling 1,000 pounds-force
K_v	Track vertical stiffness
L/V	Lateral to vertical force ratio
L_d	Deanchoring length on one side of rail cut when destressing
ℓ_{max}	Maximum permissible distance between two measurement points
l_c	Typical car length (expressed in miles)
N	Number of ties used in the 50 foot zone
N_B	Maximum number of permissible buckles per annum per section of track

n	Sampling rate
n_c	Number of cars in a typical train
n_t	Number of trains per day
n_p	Number of potential buckling events possible
$\overline{P}$	Rail force in the buckled zone
P	Longitudinal rail force
P_a	Annual probability of buckling
P_b	Probability of buckling
P_T	Relative frequency distribution of rail temperature
Q	Track weight per unit length
R	Radius of curved track
R_v	Vehicle contributes an additional load
S	Tie spacing
S_w	Shoulder width
T_{all}	Allowable rail temperature increase
T_c	Critical temperature for slow orders
T_{Bmax}	Upper critical temperature
T_{Bmin}	Lower critical temperature
T_L	Limiting temperature for train operations
T_{MAX} or T_M	Maximum rail temperature
T_N or N	Rail neutral temperature
T_M	Maximum rail temperature
T_P	Temperature at which increased lateral displacements begin to occur with small increases in rail temperature
T_R	Annual rail temperature frequency
U	Rail longitudinal displacement in adjoining section
u	Rail longitudinal displacement in buckled zone
V	Axle vertical load
V_{max}	maximum permissible speed
Vr	Reduced speed
v	Rail vertical displacement
w	Lateral deflection
w_B	Pre-buckling deflection
w_C	Post buckling displacement
w_o	Initial misalignment function
w_P	Peak lateral displacement
W_{tie}	Weight of one tie including fasteners and plates
W_{rail}	Rail weight
Y	Confidence level